Journal of Neural Transmission

Supplementum 36

JNT

A. H. Tuma, E. M. Stricker, and S. Gershon (eds.)

Advances in Neuroscience and Schizophrenia

Springer-Verlag Wien New York

Prof. A. H. Tuma
Western Psychiatric Institute and Clinic, University of Pittsburgh School of Medicine, Pittsburgh, PA, U.S.A.

Prof. E. M. Stricker
Department of Behavioral Neuroscience, University of Pittsburgh, Pittsburgh, PA, U.S.A.

Prof. S. Gershon
Western Psychiatric Institute and Clinic, University of Pittsburgh School of Medicine, Pittsburgh, PA, U.S.A.

Typeset by Best-set Typesetter Ltd, Hong Kong

Printed on acid-free paper

With 42 Figures

ISSN 0303-6995
ISBN-13:978-3-211-82347-7 e-ISBN-13:978-3-7091-9211-5
DOI: 10.1007/978-3-7091-9211-5

Preface

On May 6, 1991, the Center for Neuroscience and Schizoprenia of the University of Pittsburgh began a two-day workshop to discuss recent advances in basic neuroscience research related to schizophrenia. Schizophrenia is an extraordinarily complex disorder of brain function, which presents as a collection of phenomena that can vary both across individuals and within the same individual as a function of time, environment, and drug treatment. For these and other reasons, it has largely defied thorough analysis and understanding to date. However, we thought that a workshop on the biology of this disorder would be useful both to summarize recent progress in this area of research, and to stimulate discussion among scientific investigators and clinicians.

There are four general approaches to contemporary studies of the biological bases of shizophrenia, and each was highlighted in the presentations made during the workshop. First, specific brain sites (e.g., prefrontal cortex) have been targeted whose dysfunction are suspected to be involved in the pathogenesis of schizophrenia, and investigators have sought to learn as much as possible about the anatomy and function of these sites and their inter-relations. Second, chemically specific transmitter systems in brain (e.g., dopamine) have been targeted whose dysfunctions are suspected to be involved in the pathogenesis of shizophrenia, and investigators have sought to learn as much as possible about the anatomy and function of these systems and their inter-relations. Third, specific psychological functions (e.g., "working memory") have been targeted whose disruption is suspected of being critical to the pathogenesis of schizophrenia, and investigators have sought to learn as much as possible about their biological bases. Finally, specific techniques to measure brain function in vivo in schizophrenic patients and control subjects (e.g., brain imaging) have been targeted, and investigators have sought to understand the attributes and limitations of these techniques, and how they may best be refined and used in studies of schizophrenia.

Aside from the numerous details concerning the prefrontal cortex and basal ganglia, dopamine, memory, and brain imaging that emerged from the discussion, there was an exhilarating sense of progress in the ability of investigators to make progress. For example, it was just 20 years ago that we moved from measuring dopamine in whole brain to measuring it in whole striatum, and now we have the techniques to reveal the exquisite story of patch/matrix distribution of dopamine and its neuroanatomical connections. In addition to such advances in basic techniques, there also was an exhilarating variety of ideas being considered and pursued experimentally. This field of research evidently is alive and thriving, and there is considerable enthusiasm in the quest for insight and understanding.

A good experiment is characterized not only by the quality of question that it addresses and answers, but also by the new questions it opens for investigation. The same is true of a good workshop. The panelists have asked many specific questions in their presentations, but they also raised several general questions for later consideration. There are three that deserve emphasis. First, what is the functional significance of the basic biological observations? How is brain function influenced by the existence of the patch-matrix configuration in the striatum, for example, or the presence of intrinsic catecholaminergic neurons in prefrontal cortex? Second, what is the relevance to human cognitive function of these observations in animals? How do studies of dopamine's function as a neuromodulator, or the effects of stress on brain function, translate from animals to humans? And third, what is the relevance of these issues to schizophrenia? Specifically, what does the basal ganglia have to do with schizophrenia, and how do dysregulations in dopamine manifest themselves in symptoms of schizophrenia?

The stated goal of the workshop was to provide a forum for presentation and discussion on some of this recent progress. The hidden agenda was to provide a workshop that was especially attractive to graduate students, postdoctoral fellows, research associates, medical residents, and junior faculty, so that they might get involved in these issues relatively early in their professional development and make studies of schizophrenia their lifelong interest. In that regard, we were especially pleased that there were so many young people who registered and were in attendance at the workshop. Similarly, we are very pleased that Springer-Verlag Wien is publishing the proceedings of the workshop and thereby giving it a wider, international audience. The ultimate success of the workshop will depend on the impact it had not only on those presently involved in the field but on future generations of workers who were intrigued by the mysteries and inspired by the advances of the present work.

To those young people, we say that you should appreciate how much progress has been made in a rapidly advancing field, but you should not be overwhelmed by that progress or fear that you are getting into the field too late, after the most important issues have all been addressed and answered. In fact, you should understand that we still have not answered any of the basic questions, and that we are still closer to the beginning than to the end of this quest.

The Center for Neuroscience and Schizophrenia is an interdisciplinary center at the University of Pittsburgh comprising faculty in the Department of Behavioral Neuroscience in the Faculty of Arts and Sciences and faculty in the Department of Psychiatry in the School of Medicine. Its activities are supported by a grant from the National Institute of Mental Health (MH-45156). This workshop was co-sponsored by the Department of Psychiatry, the Department of Behavioral Neuroscience, and the Center for Neuroscience, for which we are grateful. We are also pleased to acknowledge additional financial contributions from Lilly, Sandoz, Upjohn, and Wyeth-Ayerst pharmaceutical companies. Finally, we are grateful for the contributions of all the participants and support staff to the workshop, particularly those of Jay W. Pettegrew, M.D., Theodore W. Berger, Ph.D., and Robert Y. Moore, M.D., Ph.D., who shared the chairmanship of the three sessions.

A. H. TUMA
E. M. STRICKER
S. GERSHON

Pittsburgh, May 1992

Contents

Imaging

Basal ganglia

Prefrontal cortex

Listed in Current Contents

Imaging

J Neural Transm (1992) [Suppl] 36: 3–12

Critical review of PET, SPECT and neuroreceptor studies in schizophrenia

T. F. Budinger

Lawrence Berkeley Laboratory, University of California, Berkeley, CA, U.S.A.

Summary. Hypotheses having to do with the neurochemical basis for mental disease can be tested by noninvasive measurement of flow, metabolism and neuroreceptor concentrations in various parts of the brain. The capabilities of these technologies for accurate absolute quantitation are limited by the instrument resolution and the limited statistics associated with allowable radiation doses. Nevertheless, very important new information regarding the brain function in normal and diseased states can be gleaned from the techniques of positron emission tomography (PET) and single photon emission tomography (SPECT). This chapter presents the critical problems which face the clinical researcher who endeavors to test hypotheses regarding mental illnesses using PET or SPECT. First, the quantitative capabilities of emission tomography will be discussed; then the special requirements of SPECT for attenuation correction will be reviewed. Next, the physiological principles required for inter-institutional comparisons are presented.

Resolution and quantitation

Commercial PET and SPECT instruments have the ability to give valid numbers for flow, metabolism, and receptor densities only in 10 mm sized regions of the brain. Whereas the stated resolution of a PET instrument might be 6 mm, the quantitative recovery of data in regions smaller than the thickness of the cortex (e.g. 4 mm) will be less than 50% due to the fact that the limited resolution will result in the activity being spread over a region larger than the cerebral cortex, for example (Budinger et al., 1984). This spreading or blurring of activity results in a lowering of the concentration giving rise to erroneous values for flow, metabolism, and receptor density (Fig. 1). An accurate estimate of the concentration of fluorodeoxyglucose in the cortex cannot be expected unless the instrument resolution is better than the thickness of the cerebral cortex. For example, the error in glucose metabolism can be greater that 30% if the measurements are made with an instrument with a resolution of 8 mm as shown in Fig. 2. A region of

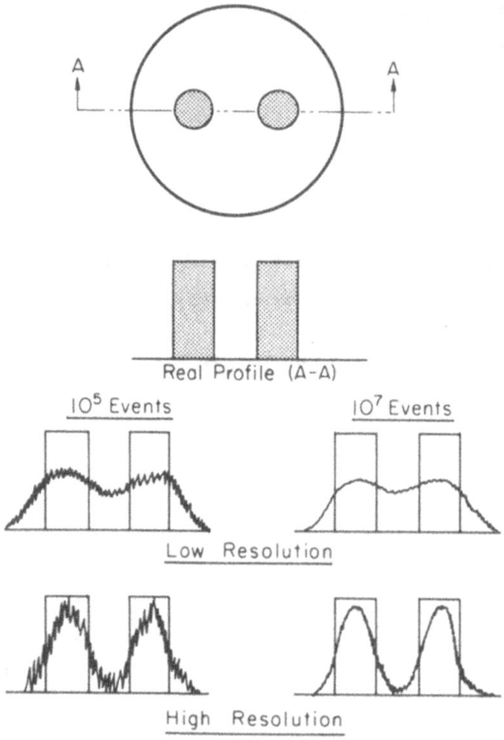

Fig. 1. Theoretical expectation of the spread of activity from an system which have a resolution comparable to the size of the actual activity distribution

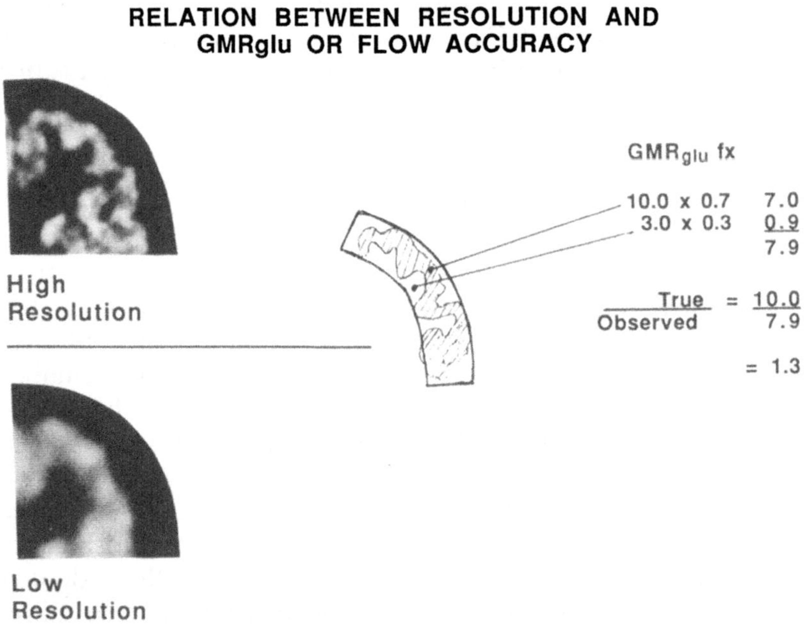

Fig. 2. Comparison between the imaging capabilities and the quantitative information for two PET instruments. Upper shows the results on the same patient as the lower image

interest over the low resolution image is shown to include both white and gray matter.

Though these facts severely limit the absolute quantitative ability of PET or SPECT, the limitations do not invalidate the techniques for measurements of relative concentrations — metabolism, receptor concentrations, and flow. When comparing one area of the cortex to another, the relative concentration is generally a reliable estimate. Indeed, if the activity adjacent to the gray matter cortical ribbon is low and the tracer is distributed uniformly in a cerebral cortical area, it is possible to correct the PET or SPECT data based on a priori anatomical information provided by NMR. In general, when examining large structures such as the striatum, the absolute quantitation will be reliable. If the resolution allows, the concentration within the putamen, for example, can be quantitated but it is not possible to achieve accurate ratios of concentration between portions of the putamen or hippocampus with the resolution of available commercial instruments.

Two areas of contemporary interest provide examples of the problem. It is of interest to evaluate the metabolism of the hippocampus in schizophrenia. Due to the limited size of the hippocampus relative to the resolution of available instruments, one must view with extreme caution even the relative concentrations of activity left to right and anterior to posterior because we can expect errors not only due to the blurring of activity but also due to partial volume contributions which are related to the slice thickness and position of the tomographic slices relative to the hippocampus.

A second aspect of research in schizophrenia which cannot be pursued easily with the present instruments is the evaluation of the balance of neuroreceptors between different parts of the cerebral cortex and between the cerebral cortex and striatum. Both the dopamine and serotonin systems exist in the cerebral cortex and hippocampal gyri. These neuroreceptors are found within thin layers of the cortex less than 1 mm thick. Considering the specific to nonspecific accumulation of ligands, the limitations of 6 mm resolution instruments in detecting concentrations in 1 mm regions and the statistical noise associated with emission tomography, it is unlikely that we will be able to explore the question of neuroreceptor relative concentration with instruments having the present resolution.

Let us examine the determinants of resolution for a PET device. The general concept of PET detection shown in Fig. 3 shows that the first determinant of resolution is the detector size. Because of the geometry of the detection of annihilation photons at 180 degrees, the geometric resolution is ½ the detector thickness. Thus, for a detector of 3 mm width, we expect 1.5 mm resolution. But there are two other physical factors which limit resolution (Fig. 3). The angulation error of 0.25 degrees at the time of photon production through the annihilation of a positron with an electron can cause a error of a few mm in the reconstructed image. The positron leaves the nucleus with kinetic energy which means it travels a short distance in tissue and this range effect can seriously affect resolution if the

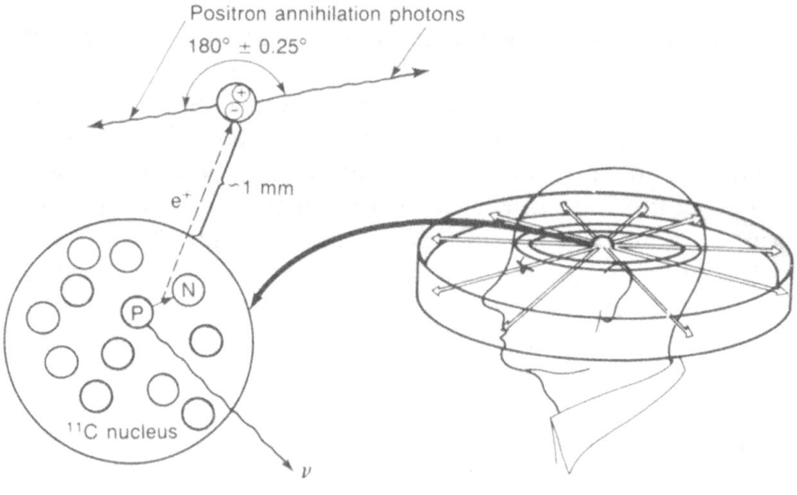

Fig. 3. Positron emission tomography concept

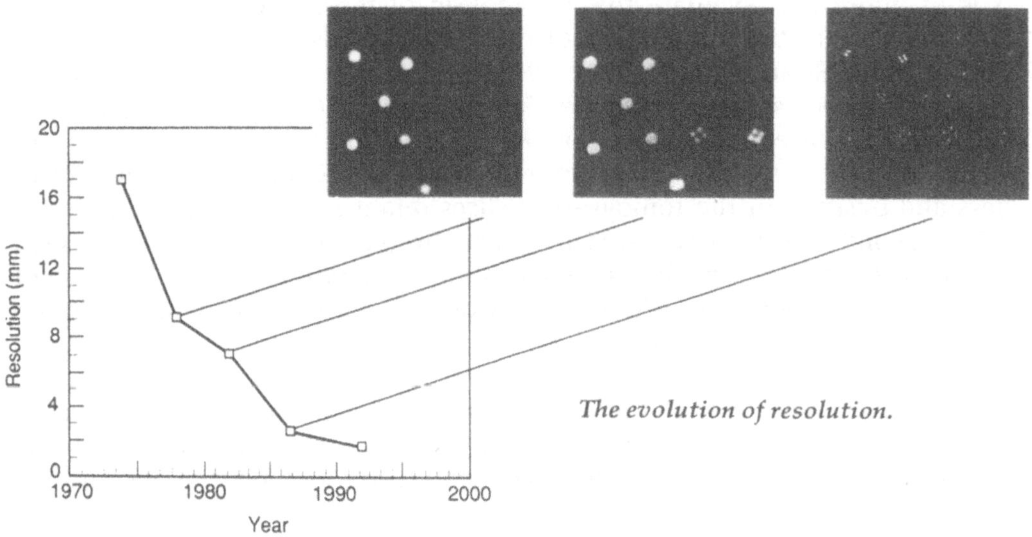

Fig. 4. Factors affecting resolution

kinetic energy is high as is the case for some positron emitters (e.g. iodine-122, iodine-124, rubidium-82); however, this effect is not as serious as some have speculated as it results in a resolution deterioration of less than 1 mm (Fig. 4). Important image resolution over the last 15 years is shown in Fig. 5. (Budinger, 1990). The Donner 600-crystal tomograph was built to demonstrate the possibilities of achieving 2.6 mm resolution using 3 mm crystals. Results from this instrument in the measurement of deoxyglucose concentration in Alzheimer's are shown in Fig. 6.

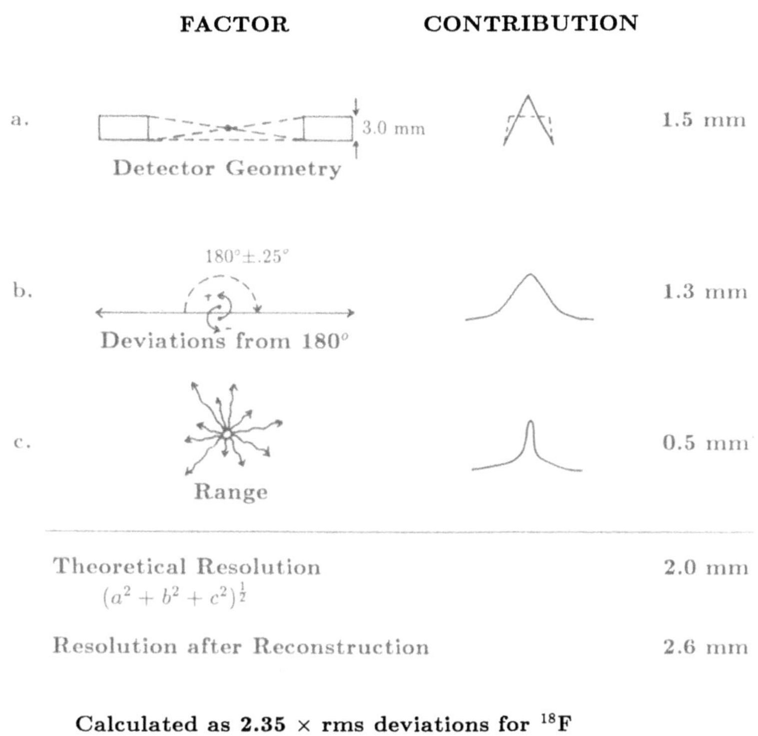

Fig. 5. Improvement in resolution over the past 15 years. Arrow indicates point sources 4 mm apart

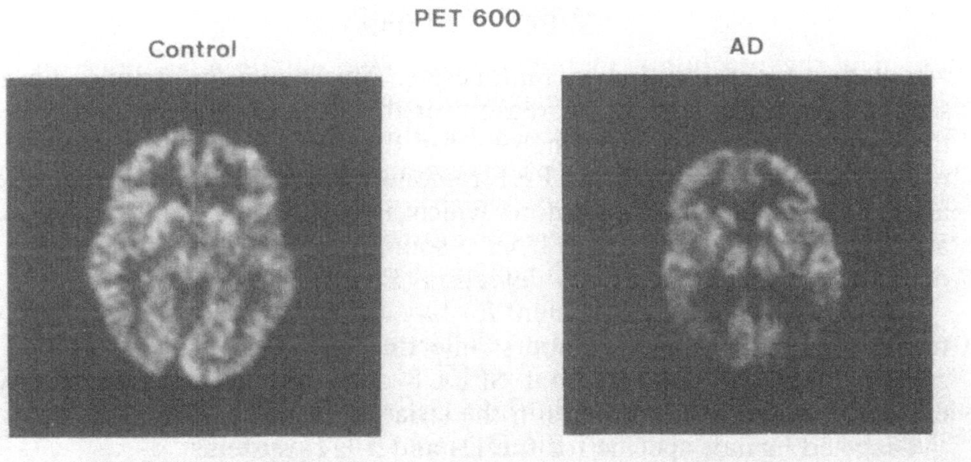

Fig. 6. PET images of a control and Alzheimer's patient

It is now believed to be possible to achieve 2 mm resolution using a new detector design (Moses et al., 1992). However, even with an instrument with a resolution of 2 mm, we would not expect an accurate concentration value for neuroreceptors because the neuroreceptors might reside in layers thinner than 1 mm. The requirement for increased sensitivity in order to achieve sufficient statistics to give reliable data with 2 mm resolution can be satisfied by a new design approach which involves new detectors just now under development and time-of-flight implementation to improve the signal-to-noise.

Sensitivity

Sensitivity is defined as the number of detected good events of the tomographic system (resolution) per a given source activity in an attenuating medium. The sensitivity of a PET system is dependent on the detector size and efficiency. A typical sensitivity of a commercial PET system is 15,000 events per second per μCi in a 20 cm diameter attenuator for 1 cm axial resolution. This rather lengthy characterization of sensitivity is needed in order to be able to compare one system to another and to be able to predict feasibility of new types of quantitative studies.

The major determinant of sensitivity of SPECT systems is the amount of detector material surrounding (usually how many cameras) the subject and the resolution which is provided by the collimator. The better the resolution, the lower the sensitivity.

SPECT systems have less sensitivity than PET by the factor given in the equation below:

$$\frac{\mathrm{PET}}{\mathrm{SPECT}} = \frac{150\,(\mathrm{mm})}{\mathrm{d}\,(\mathrm{mm})} \tag{1}$$

where d is the resolution in millimeters. This equation assumes an axial section 1 cm thick. The major reason for this difference in sensitivity lies in the need to collimate the photons for the SPECT single photon radionuclides. The collimation for PET radionuclides in given by the electronic acceptance of only those photons which arrive in time coincidence — the assumption is that they arose from a source located on a line between the detectors which registered the detection (Fig. 4).

Not only is SPECT convenient for blood flow studies wherein the resolution is adequate to give numbers reflecting blood flow in one part of the cortex relative to another, but SPECT does provide an opportunity to evaluate the receptor activities in the striatum because of the availability of ^{123}I-labeled ligands specific for the D_2 and 5-HT_2 systems.

Attenuation compensation

A critical aspect of quantitative SPECT which distinguishes this technique as potentially less quantitative than PET is the need for special manipulations of the data during reconstruction. Both in PET and SPECT, the photons are attenuated by the tissues between the detectors and the source.

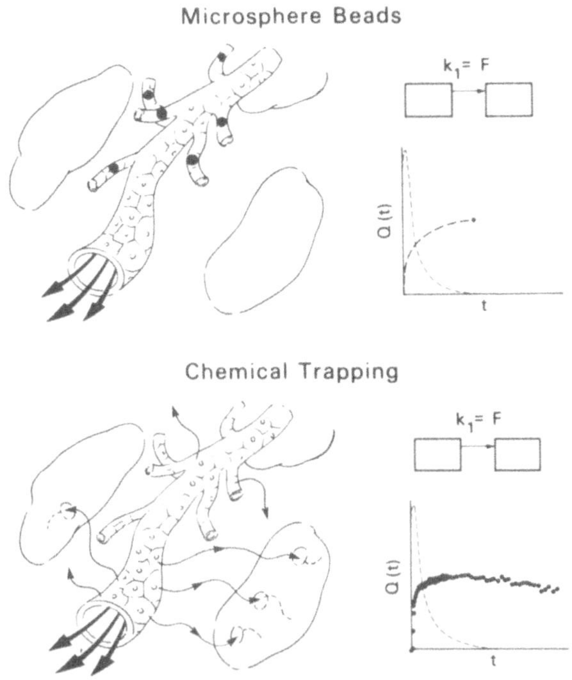

Fig. 7. Reduction of the reconstruction activity at the center of the head can be expected if attenuation correction is not applied in SPECT

The attenuation correction is easy to apply in PET and much more difficult for SPECT. The problem has not been so much in the mathematics, but in the commercial implementation of fast, yet accurate methods. Both computer memory and computational times have been limiting in the past. The problem as illustrated in Fig. 7 is the proper use of attenuation correction mathematics. For the usual radionuclides the head attenuation results in a lowering of the apparent concentration by as much as 300%.

Quantitative interinstitutional and logitudinal studies

Finally, the use of practical models which will allow the quantitative comparison between patients, and from time to time in the same patient, demands some measure of the arterial input function which is usually determined by drawing arterial blood every 5–10 seconds. The time rate of change of arterial concentration can also be determined from tomographic data obtained over the heart using dynamic PET or SPECT. The basic model for quantitating the uptake of a tracer which does not wash out of the brain is:

$$FE = Q(T) \Big/ \int_0^T A(t)dt \qquad (2)$$

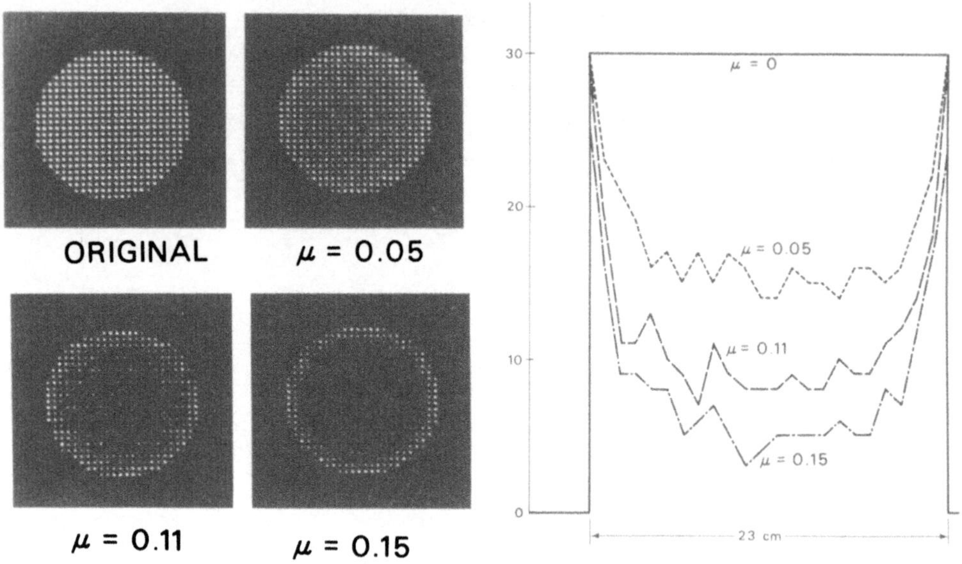

Fig. 8. The correspondence between microspheres method of measuring flow and the use of radiopharmaceuticals which are trapped in the tissues (see Fig. 9 for tracers which are not trapped)

where F is flow, E is extraction, Q is the amount in the tissue at some time, T, and A is the arterial concentration (Budinger and Kuesman, 1985). We assume that the arterial concentration is measured from an arterial line or determined from dynamic tomographic data taken over the heart during the initial few minutes after the injection. Thus, in this equation Q(T) is spatial distribution of concentration made by moving the head into the tomograph at time T. Note that for microspheres or ligands that behave like microspheres, E in the above equation is 1 (Fig. 8).

More complex models involving interchange between extravascular compartments and receptor systems as well as interchange between extravascular compartments and nonspecific accumulation sites generally require data collection strategies which place severe demands on patients, control subjects and statistical limits of the data, particularly if information from small regions is being sought.

Using a model as simple as equation 2 or a slightly more complex model for the partly diffusible tracer (Fig. 9) wherein a wash-out (k_2) is significant, reasonably good quantitative values can be determined. In this case flow in (k_1) and wash-out (k_2) are derived from fitting the data from each subregion of the brain to the equation.

$$Q(T) = (k_1) \int_0^T dtA(t)e^{-k_2(\tau - t)} \quad (3)$$

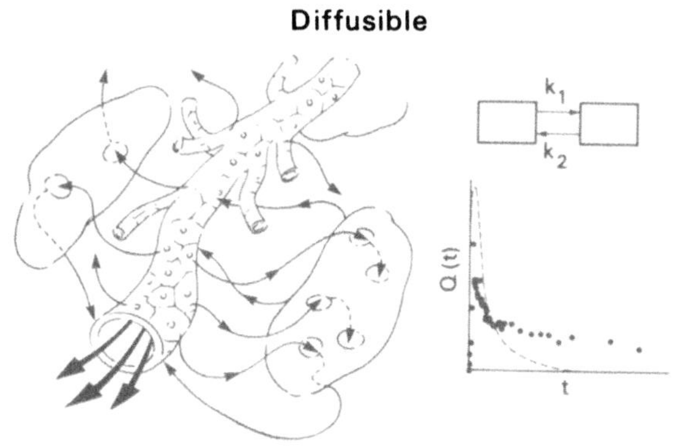

Fig. 9. The two-compartment model for a diffusible tracer such as $^{15}O\text{-}H_2O$

In the case of equation 3, Q must be measured simultaneously with A(t) thus requiring arterial sampling by catheter or yet to be developed instrumentation.

Activation studies

Finally, much of the work currently under way using activation analysis with PET and $H_2{}^{15}O$ relies on the phenomenon modeled in equation 3. Most of the information being derived on the spatial localization of cognitive processes evaluate brain image information collected by integrating Q(t) with the assumption that any changes of the arterial activity behavior A(t) will be normalized by the process of rationing or subtracting the integrated images for the conditions of control vs. activation. An alternate approach using a non-diffusible flow tracer (^{122}I HIPDM; $T_{1/2} = 3.6\,\text{min}$) and the model of equation 2 can give accurate results without the need to measure the input function if relative differences are being sought.

Conclusion

Major advances in the use of noninvasive techniques in the study of schizophrenia can be made if the restrictions of resolution and statistics are kept in mind when analyzing data which has been properly corrected for attenuation and scatter. The present resolution of commercial instruments prohibits accurate measurements of the relative concentrations of neuroreceptors between the cortex and mid-brain gray matter. The expectations for very high resolution PET instruments can be realized to the extent that structures as small as the locus ceruleus and inner vs. outer cerebral cortex can be evaluated. This will require 2 mm resolution and image acquisition times in excess of 15 min. In the interim, studies which report flow, metabolism and neuroreceptor concentrations in mental disorders vs. normal controls will be improved in their contribution if the limitations of resolution are addressed frontally.

Some cautions regarding inferences from radiopharmaceuticals

The importance of specificity of the ligand is made clear when examining the cross specificity of compounds such as methylspiperone to both the 5-HT_2 and the dopamine D_2 system. On the other hand, a highly specific ligand for the D_2 system such as raclopride might not be an ideal tracer of neuroreceptors because the low affinity ligand might be replaced in vivo by endogenous compounds such as dopamine itself. Thus, the evaluation of the D_2 system requires the careful attention to the affinity of the ligand as well as its specificity.

A parallel argument can be made for selection of flow tracers. Whereas, some cerebral blood flow tracers are readily available, they might have a metabolic component which will lead to inaccurate flow results.

References

Budinger TF (1990) Advances in emission tomography instrumentation: Quo Vadis? J Nucl Med 31:628–631

Budinger TF, Huesman RH (1985) Ten percepts for quantitative data acquisition and analysis. Circulation 72:IV-53–IV-62

Budinger TF, Derenzo SE, Huesman RH (1984) Instrumentation for positron emission tomography. Ann Neurol 15:§35–§43

Moses WW, Derenzo SE, Budinger TF (1992) Design for a high-rate, high resolution PET module using room-temperature silicon photodiodes for crystal identification. J Nucl Med 33

Author's address: Dr. T. F. Budinger, The Lawrence Berkeley Laboratory, University of California, MS 55–127, Berkeley, CA 94720, U.S.A.

J Neural Transm (1992) [Suppl] 36: 13–22

MRI and cognitive behavioral function in schizophrenia

R. E. Gur

Department of Psychiatry, University of Pennsylvania, Philadelphia, PA, U.S.A.

Summary. There is active research applying MRI to the study of brain anatomy in schizophrenia. As the technology improves and measurements are refined, this method, in combination with behavioral measures, is likely to contribute to the understanding of brain function in schizophrenia. The heterogeneity of schizophrenia challenges investigators to integrate anatomic and behavioral parameters with physiology. MRI techniques have already been applied to obtain quantitative imaging of cerebral blood flow and holds promise for combined MRI and neurobehavioral studies.

Introduction

The severity of behavioral abnormalities in schizophrenia has led many investigators to persist in search of neural dysfunction. Neuroimaging has provided unprecedented opportunities for a systematic examination of neural substrates for behavior and can be used as a powerful aid in understanding the pathophysiology of schizophrenia. In this review the MRI and the neuropsychological literature on schizophrenia will be examined and future directions suggested.

MRI studies in schizophrenia

The application of MRI to the study of the neuroanatomy of schizophrenia is a relatively recent addition to more than a decade of CT research on brain structure in schizophrenia. The CT literature has been reviewed by Coffman (1989). Most studies have suggested increased ventricular-brain ratio (VBR). The focus on VBR seems related to the ease of visualization of ventricles and the ability to obtain reliable linear and planimetric measures. Furthermore, CT is not optimal for measuring sulci.

Studies which have examined sulci, measured specific sulci or applied global rating scales. For example, Weinberger et al. (1979) measured linearly the largest sulci and reported enlargement in 32% of schizophrenics. Johnstone et al. (1978) and Iacono et al. (1988) found no sulcal enlargement in patients using global ratings. The technical limitations of CT in obtaining sulcal measures were addressed in Pfefferbaum et al. (1986). In a more recent study, they used a semi-automated computerized approach for volumetric measurements. Diffuse cerebral atrophy distinguished schizophrenics from controls. Sulcal, more than ventricular enlargement, was implicated (Pfefferbaum et al., 1988). MRI is a more sensitive technique for examining sulcal changes and permits visualization of temporal and limbic structures implicated in neuropathologic studies of schizophrenia (Bogerts et al., 1985; Brown et al., 1986; Arnold et al., 1991). The advantages of MRI also include better resolution, higher sensitivity to soft tissue contrast, multiplanar imaging capabilities, no bone artifact, feasibility for combined study of structure and metabolic function, and no ionizing radiation.

There is a rapidly growing literature on MRI studies in schizophrenia. Smith et al. (1985) studied 23 patients and 17 nonpsychotic volunteers with a 0.3 T device. Linear and area measurements of several brain structures revealed no differences between groups. Higher T1 relaxation time in some brain regions of schizophrenics was noted. These results were maintained with a somewhat larger sample (Smith et al., 1987). Mathew et al. (1985) evaluated 18 patients and 18 controls with a 0.5 T scanner, comparing the surface areas of the septum pellucidum, midsagittal brain slice, and the ratio between the two. Patients had larger septa and septo-brain ratios. Measurements of the cerebellar vermis and height of the fourth ventricle did not distinguish 12 schizophrenics from 12 controls (Mathew and Partain, 1985). Patients ($n = 27$) did not differ from controls ($n = 12$) on clinical reading of increased signal intensity in periventricular spin echo images (Johnstone et al., 1986). Andreasen et al. (1986) evaluated 38 schizophrenics and 48 controls with a 0.5 T instrument, reporting that on midline sagittal slices patients had smaller frontal lobes as well as smaller cerebrums and craniums. Since normals did not match schizophrenics sociodemographically, a second sample of 26 patients was compared to 27 matched controls, revealing no differences between the groups in these measures (Andreasen, 1989). The original Iowa sample was examined for callosal thickness as it relates to gender, and schizophrenic women were found to have an increase in callosal middle and anterior thickness compared to controls (Nasrallah et al., 1986). Using a 0.08 T scanner, Besson et al. (1987) compared 23 schizophrenics to 15 controls. While there was no diagnosis effect, T1 was reportedly increased in some brain regions for subgroups of patients. Kelsoe et al. (1988) compared 24 schizophrenics to 14 controls with a 0.5 T scanner using midsagittal and coronal sections. Linear, area, and signal intensity were examined in nine regions. In four regions volumetric estimates were obtained. The third ventricle was reported to be 73% larger in area, and the lateral ventricles 63% larger in volume in patients. No other differences were noted between the groups. Increased VBR was reported for 12 schizophrenics, compared to 12 controls, as well as a smaller corpus callosum (Rossi et al., 1989). We have compared 42 patients with 43 controls examining brain and CSF volumetrically (Gur et al., 1991). Patients did not differ from controls in whole brain volume, but had higher whole brain CSF volume than normals and higher ratios of ventricular and sulcal CSF to cranial volume.

The earlier attempts to apply MRI have not used the technology to its full capability and can be improved with higher resolution scanners, increased sophistication of image analysis, and larger samples. These permit more precise examination of regional abnormalities in brain structure. The emphasis in neuroanatomic research has shifted to examination of structures implicated in the pathophysiology of schizophrenia. A number of studies measured the size of the temporal lobe and related limbic structures. Suddath et al. (1990) studied 15 sets of monozygotic twins discordant for schizophrenia. Smaller anterior hippocampi and enlarged lateral and third ventricles were evident in most of the affected twins when compared to the unaffected ones. Bogerts et al. (1990) measured the volume of the hippocampus-amygdala complex and adjacent temporal horns in 34 patients with schizophrenia assessed during the first episode of illness and compared to 25 controls. T1 weighted contiguous coronal sections were used. Abnormalities were seen in a subsample of patients with males having a smaller left hippocampus. Enlargement of the left temporal horn was seen in men and women. DeLisi et al. (1991) compared first-episode (n = 30) to chronic patients (n = 15) and to neurological controls (n = 20). Only chronic patients had reduced temporal lobe size, greater on the left side. Barta and colleagues (1990) reported smaller volume of the superior temporal gyrus and the left amygdala in patients. The anatomic changes were related to severity of auditory hallucinations. Rossi et al. (1990) reported decreased volume in the left temporal lobe of 17 patients compared to 13 controls. Young et al. (1991) studied 31 patients and 33 controls and found that the temporal lobe was smaller on the left than the right in both groups. A series of studies reported abnormalities of the corpus callosum in schizophrenia (e.g., Casanova et al., 1990; Raine et al., 1990). Thus, MRI has built upon a large number of studies with CT suggesting that some abnormalities in brain anatomy and schizophrenia are evident but that the yield might be greater in examining specific brain structures rather than ventricles alone.

Studies are now examining specific regions of interest. While this is an important step there are a number of considerations to keep in mind. Each laboratory has established its own procedures for determination of anatomic boundaries and data on reliability, when provided, are limited. Another important issue is that of specificity of findings. As studies of the whole brain have suggested, patients with schizophrenia may have generalized atrophy and this makes reports of regional atrophy difficult to interpret unless it is shown that atrophy in a specific region significantly exceeds atrophy in other "control" regions. Finally, few studies have related anatomic findings to behavioral measures.

Neurobehavioral studies

Cognitive deficits in schizophrenia have been observed and documented. The neuropsychiatric perspective of the disorder has sharpened hypotheses which tie aberrant behavior with regional brain dysfunction. This growing literature is based on neuropsychological theories and practice (see Gur et al., 1990 for review). Anterior brain systems, particularly the frontal lobe, have been implicated by observed deficits in executive functions, abstraction and mental flexibility on tests such as the Wisconsin Card Sorting Test and the Halstead Category Test (Flor-Henry and Yeudall, 1979; Flor-Henry et al., 1983; Weinberger et al., 1986). Disturbed laterality with involvement of the left hemisphere has been suggested by deficits in cognitive abilities which are commonly subserved by the left hemisphere. Thus, low VIQ relative to PIQ and impairment on language tests (Flor-Henry and Yeudall, 1979; Newlin et al., 1981 for reviews) support this hypothesis. Subcortical and cortical dysfunction has been related to studies of attention and information processing (Mirsky, 1977, 1986).

The extensive interconnectivity between brain structures indicates a multi-system deficit rather than discrete involvement of brain structures. This is compatible with the behavioral presentation of schizophrenia where symptoms are diverse and cognitive deficits diffuse. For example, subcortical systems connected to the frontal lobe are potential sites for disruption of attentional processes via impaired selective gaiting of information (Skinner and Lindsley, 1973). Mesulam (1985) has proposed a neural network for attention, which involves parietal, temporal and frontal cortex as well as the cingulate gyrus and subcortical "limbic" regions.

There are many methodological problems in psychometric studies of schizophrenia (Heaton and Crowley, 1981; Cassens et al., 1990). In earlier studies, the conceptual basis for applying neurobehavioral methods was an attempt to separate patients with "functional" schizophrenia from those with organic brain dysfunction. While the general finding has been that the level of performance of schizophrenics falls between that of patients with known structural brain lesions and controls, patients were not well characterized and therefore not permitting a careful linking between phenomenology and measured behaviors. Furthermore, the interpretation of the obtained deficits requires the formation and analysis of quantitative profiles and the mapping of these profiles onto regional brain systems. Since multiple brain systems are involved in schizophrenia, the question becomes how to test which function is differentially impaired (Chapman and Chapman, 1989), and what are the implications of such differential impairment to hypotheses on regional brain dysfunction (Gur et al., 1990).

We have applied a battery which measures a range of behavioral functions and implicated brain regions. From this battery it is possible to assess regional disturbances and test specific hypothesis such as frontal relative to temporal lobe impairment. The frontal lobe hypothesis will predict that patients are more impaired on the Wisconsin Card Sorting Test than on the memory and learning tests, whereas the temporal lobe hypothesis will predict the reverse. In a sample of 36 patients and 36 normal controls (Saykin et al., 1991), we found that patients were diffusely impaired both in abstraction (WCS) and memory tests. However, the impairment in memory and learning was significantly worse than the impairment on the WCS. This effect, in support of temporal lobe deficit in schizophrenia, remained when variables which might affect performance were considered in the statistical analysis. Thus, it existed for both men and women, across the age range (18–45), for all educational levels, and for both highly cooperative and less cooperative patients.

There is need for a systematic evaluation of the combined set of neuropsychological data in relation to current theories of brain behavior regulation. This will permit the identification of impaired neural networks. The nature of the brain disease we expect to find in schizophrenia is not a focal lesion with circumscribed boundaries and effects, but more likely involving neurotransmitter systems with distributed physiologic effects. Given that, it would be a mistake to find an area of greatest abnormality and declare it as the site of schizophrenia. Rather, we will have to continue and search systematically for networks. This requires the joining of neuropsychological test scores with hypotheses linking them to the integrity of all regions of interest.

Relation of neuroanatomic to behavioral measures

While neuroanatomic studies are important for improved understanding of normal and pathological brain function, their integration with clinical, neurobehavioral and physiologic data is essential for the study of regional brain function in schizophrenia (Gur, 1985). Many CT studies have examined the relation between VBR and clinical features (e.g., Andreasen et al., 1982; Johnstone et al., 1978; Williams et al., 1985). The results are inconclusive (e.g., Owens et al., 1985; Pfefferbaum et al., 1988). This can be expected in a heterogeneous disorder when small samples are studied with different methodologies. Establishing a link between the anatomic and clinical findings is requisite for exploiting the potential diagnostic and prognostic utility of the neuroimaging techniques. The question of when in the course of illness do structural abnormalities appear has been addressed by evaluating patients with schizophreniform disorder (Weinberger et al., 1982), schizophrenia spectrum disorder (Schulz et al., 1983), and patients with new onset of psychosis (Iacono et al., 1988; DeLisi et al., 1991). These studies suggest that differences between schizophrenics and normals may exist at onset. Weinberger (1987) proposed that a static lesion accounts for the neuroanatomic abnormalities. Few studies have addressed the stability of ventricular enlargement in schizophrenia. Illowsky et al. (1988) rescanned 15 chronic schizophrenics after 7–9 years of continuous illness. No changes were noted in VBR or frontal atrophy scores. A preliminary followup study by Nasrallah et al. (1986) on 11 patients, on different scanners 3 years apart, also showed no consistent change. Although these studies appear to support a static lesion, the sensitivity of the measures does not permit ruling out progression.

Of the many recent studies on cognitive performance in schizophrenia, some have examined the relationship between neuroanatomic measures and cognitive functioning. Golden et al. (1980) reported correlations (.33 to .58) between 8 of 14 Luria-Nebraska scales and VBR. In a replication (Golden et al., 1982), all 15 patients with enlarged ventricles were classified as brain-damaged, and so were 36% of patients with normal ventricles. Donnelly et al. (1980) found an association between impairment on the Halstead-Reitan battery and increased VBR in 15 chronic schizophrenics. However, Carr and Wedding (1984) reported that all chronic patients, regardless of ventricular size, scored in the "brain-damaged" range with this battery. Kemali et al. (1985) found a subsample (7 out of 33) of schizophrenics with increased (>2 SD) VBR who had poorer performance on some scales of the Luria-Nebraska battery and the WAIS. For the whole sample there were no correlations between VBR and any of the measures. Andreasen et al.'s (1986) MRI study reported correlations between visual memory and measures of frontal function and residual cranial size in schizophrenics. Pfefferbaum et al. (1988) found correlations between the Luria-Nebraska global impairment score and any of the anatomic measures.

Establishing relations between neurobehavioral measures and neuroanatomy is a challenging endeavor. It requires the coupling of two different sets of variables, each with its own unreliability, and which are each highly intercorrelated. The studies so far are preliminary, since larger samples and a comprehensive and systematic approach to behavioral assessment are needed for linking behavioral dimensions to neuroanatomy.

It has been recognized that neuroanatomic abnormalities are unlikely to be the sole cause of brain dysfunction in schizophrenia, and physiologic abnormalities have been suggested. Advancing the field requires application of MRI with both behavioral and physiological measures. Each of these domains presents many methodological issues and the integration is a challenge. The technology is there, however, for moving the field ahead.

Acknowledgements

This investigation was supported by National Institute of Mental Health grant MH-42191, by Mental Health Clinical Research Center grant MH-43880, and by Research Scientist Development Award MH-00586 (Dr. R.E. Gur). We thank Ms. M. Taleff, MS, for assistance.

References

Andreasen NC (1989) Nuclear magnetic resonance imaging. In: Andreasen NC (ed) Brain imaging: applications in psychiatry. American Psychiatric Press, Washington DC, pp 67–121

Andreasen NC, Olsen SA, Dennert JW, Smith MR (1982) Ventricular enlargement in schizophrenia: relationship to positive and negative symptoms. Am J Psychiatry 139: 297–302

Andreasen NC, Nasrallah HA, Dunn V, Olson SC, Grove WM, Ehrhard JC, Coffman JA, Crossett JH (1986) Structural abnormalities in the frontal system in schizophrenia: a magnetic resonance imaging study. Arch Gen Psychiatry 43: 136–144

Arnold SE, Hyman BT, Van Hoesen G, Damasio AR (1991) Some cyto-architectural abnormalities of the entorhinal cortex in schizophrenia. Arch Gen Psychiatry 48: 625–632

Barta PE, Pearlson GD, Powers RE, Richards SS, Tune LE (1990) Auditory hallucinations and smaller superior temporal gyral volume in schizophrenia. Am J Psychiatry 147: 1457–1462

Besson JAO, Corrigan FM, Cherryman GR, Smith FW (1987) Nuclear magnetic resonance brain imaging in chronic schizophrenia. Br J Psychiatry 150: 161–163

Bogerts B, Meertz E, Schonfeldt-Bausch R (1985) Basal ganglia and limbic system pathology in schizophrenia: a morphometric study of brain volume and shrinkage. Arch Gen Psychiatry 42: 784–791

Bogerts B, Ashtari M, Degreef G, Alvir JM, Bilder RM, Lieberman JA (1990) Reduced temporal limbic structure volumes on magnetic resonance images in first episode schizophrenia. Psychiatr Res 35: 1–13

Brown R, Colter N, Corsellis JA, Crow TJ, Frith CD, Jagoe R, Johnstone EC, Marsh L (1986) Postmortem evidence of structural brain changes in schizophrenia: differences in brain weight, temporal horn area, and parahippocampal gyrus compared with affective disorder. Arch Gen Psychiatry 43: 36–42

Carr EG, Wedding D (1984) Neuropsychological assessment of cerebral ventricular size in chronic schizophrenics. Int J Clin Neuropsychol 6: 106–111

Casanova MF, Sanders RD, Goldberg TE, Bigelow LB, Christison G, Torrey EF, Weinberger DR (1990) Morphometry of the corpus callosum in monozygotic twins discordant for schizophrenia: a magnetic resonance imaging study. J Neurol Neurosurg Psychiatry 53: 416–421

Cassens G, Inglis AK, Appelbaum PS, Gutheil TG (1990) Neuroleptics: effects on neuropsychological function in chronic schizophrenic patients. Schizophr Bull 16: 477–499

Chapman LJ, Chapman JP (1989) Strategies for resolving the heterogeneity of schizophrenics and their relatives using cognitive measures. J Abnorm Psychol 98: 357–366

Coffman JA (1989) Computed tomography in psychiatry. In: Andreasen NC (ed) Brain imaging: applications in psychiatry. American Psychiatric Press, Washington DC, pp 1–65

DeLisi LE, Hoff AL, Schwartz JE, Shields GW, Halthore SN, Gupta SM, Henn FA, Anand AK (1991) Brain morphology in first-episode schizophrenic-like psychotic patients: a quantitative magnetic resonance imaging study. Biol Psychiatry 29: 159–175

Donnelly EF, Weinberger DR, Waldman IN, Wyatt RJ (1980) Cognitive impairment associated with morphological brain abnormalities on computed tomography in chronic schizophrenic patients. J Nerv Ment Dis 168: 305–308

Flor-Henry P, Yeudall LT (1979) Neuropsychological investigations of schizophrenia and manic-depressive psychoses. In: Gruzelier J, Flor-Henry P (eds) Hemispheric asymmetries of function in psychopathology. Elsevier/North-Holland, Amsterdam

Flor-Henry P, Fromm-Auch D, Schopflocher D (1983) Neuropsychological dimensions in psychopathology. In: Laterality and psychopathology. Elsevier Bio-Medical Press, Amsterdam
Golden CJ, Moses JA, Zelazowski R, Graber B, Zatz LM, Horvath TB, Berger PA (1980) Cerebral ventricular size and neuropsychological impairment in young chronic schizophrenics: measurement by the standardized Luria-Nebraska neuropsychological battery. Arch Gen Psychiatry 37: 619–623
Golden CJ, MacInnes WD, Ariel RN, Ruedrich SL, Chu C, Coffman JA, Graber B, Bloch S (1982) Cross-validation of the ability of the Luria-Nebraska neuropsychological battery to differentiate chronic schizophrenics with and without ventricular enlargement. J Consult Clin Psychol 50: 87–95
Gur RC (1985) Measurement of regional brain physiology in humans: early applications in behavioral neurology. In: Mesulam M-M (ed) Principles of behavioral neurology. Davis, Philadelphia
Gur RE, Gur RC, Saykin AJ (1990) Neurobehavioral studies in schizophrenia: implications for regional brain dysfunction. Schizophr Bull 16: 445–451
Gur RE, Mozley PD, Resnick SM, Shtasel D, Kohn M, Zimmerman R, Herman G, Atlas S, Grossman R, Erwin R, Gur RC (1991) Magnetic resonance imaging in schizophrenia. I. Volumetric analysis of brain and cerebrospinal fluid. Arch Gen Psychiatry 48: 407–412
Heaton RK, Crowley TJ (1981) Effects of psychiatric disorders and their somatic treatments on neuropsychological test results. In: Filskov SB, Boll TJ (eds) Handbook of clinical neuropsychology. Wiley-Interscience, New York
Iacono WG, Smith GN, Moreau M, Beiser M, Fleming JA, Lin TY, Flak B (1988) Ventricular and sulcal size at the onset of psychosis. Am J Psychiatry 145: 820–824
Illowsky BP, Juliano DM, Bigelow LB, Weinberger DR (1988) Stability of CT scan findings in schizophrenia: results of an 8 year follow-up study. J Neurol Neurosurg Psychiatry 51: 209–213
Johnstone EC, Crow TJ, Frith CD, Stevens M, Kreel L, Husband J (1978) The dementia of dementia praecox. Acta Psychiatr Scand 57: 305–324
Johnstone EC, Crow TJ, Macmillan JF, Owens DG, Bydder GM, Steiner RE (1986) A magnetic resonance study of early schizophrenia. J Neurol Neurosurg Psychiatry 49: 136–139
Kelsoe JR, Cadet JL, Pickar D, Weinberger DR (1988) Quantitative neuroanatomy in schizophrenia: a controlled magnetic resonance imaging study. Arch Gen Psychiatry 45: 533–541
Kemali D, Maj M, Galderisi S, Ariano MG, Cesarelli M, Milici N, Salvati A, Valente A, Volpe M (1985) Clinical and neuropsychological correlates of cerebral ventricular enlargement in schizophrenia. J Psychiatr Res 19: 587–596
Mathew RJ, Partain CL (1985) Midsagittal sections of the cerebellar vermis and fourth ventricle obtained with magnetic resonance imaging of schizophrenic patients. Am J Psychiatry 142: 970–971
Mathew RJ, Partain CL, Prakash R, Kulkarni MV, Logan TP, Wilson WH (1985) A study of the septum pellucidum and corpus callosum in schizophrenia with MR imaging. Acta Psychiatr Scand 72: 414–421
Mesulam M-M (1985) Principles of behavioral neurology. Davis, Philadelphia
Mirsky AF (1977) Behavioral and electrographic measures of attention in children at risk for schizophrenia. J Abnorm Psychol 86: 27–33
Mirsky AF (1986) From Worcester to Haifa: fifty years of attention research in neuropsychiatry. Paper presented at the American Psychological Association, Washington, DC
Nasrallah HA, Andreasen NC, Coffman JA, Olson SC, Dunn VD, Ehrhardt JC, Chapman SM (1986) A controlled magnetic resonance imaging study of corpus callosum thickness in schizophrenia. Biol Psychiatry 21: 274–282

Newlin DB, Carpenter B, Golden CJ (1981) Hemispheric asymmetries in schizophrenia. Biol Psychiatry 16: 561–582

Owens DGC, Johnstone EC, Crow TJ, Frith CD, Jagoe JR, Kreel L (1985) Lateral ventricular size in schizophrenia: relationship to the disease process and its clinical manifestations. Psychol Med 15: 27–41

Pfefferbaum A, Zatz LM, Jernigan TL (1986) Computer-interactive method for quantifying cerebrospinal fluid and tissue in brain CT scans: effects of aging. J Comput Assist Tomogr 10: 571–578

Pfefferbaum A, Zipursky RB, Lim KO, Zatz LM, Stahl SM, Jernigan TL (1988) Computed tomographic evidence for generalized sulcal and ventricular enlargement in schizophrenia. Arch Gen Psychiatry 45: 633–640

Raine A, Harrison GN, Reynolds GP, Sheard C, Cooper JE, Medley I (1990) Structural and functional characteristics of the corpus callosum in schizophrenics, psychiatric controls, and normal controls. A magnetic resonance imaging and neuropsychological evaluation. Arch Gen Psychiatry 47: 1060–1064

Rossi A, Stratta P, Gallucci M, Passariello R, Casacchia M (1989) Quantification of corpus callosum and ventricles in schizophrenia with nuclear magnetic resonance imaging: a pilot study. Am J Psychiatry 146: 99–101

Rossi A, Stratta P, D'Albenzio L, Tartaro A, Schiazza G, di Michele V, Bolino F, Casacchia M (1990) Reduced temporal lobe areas in schizophrenia: preliminary evidences from a controlled multiplanar magnetic resonance imaging study. Biol Psychiatry 27: 61–68

Saykin AJ, Gur RC, Gur RE, Mozley PD, Mozley LH, Resnick SM, Kester DB, Stafiniak P (1991) Neuropsychological function in schizophrenia: selective impairment in memory and learning. Arch Gen Psychiatry 48: 618–624

Schulz SC, Koller MM, Kishore PR, Hammer RM, Gehl JJ, Friedel RO (1983) Ventricular enlargement in teenage patients with schizophrenia spectrum disorder. Am J Psychiatry 140: 1592–1595

Skinner JE, Lindsley DB (1973) The nonspecific mediothalamic-frontocortical system: its influence on electrocortical activity and behavior. In: Pribram KH, Luria AR (eds) Psychophysiology of the frontal lobes. Academic Press, New York, pp 185–234

Smith RC, Baumgartner R, Calderon M, Affas A, Ravichandran GK, Peters ID (1985) Magnetic resonance imaging studies of schizophrenia. Psychopharmacol Bull 21: 588–594

Smith RC, Baumgartner R, Calderon M (1987) Magnetic resonance imaging studies of the brains of schizophrenic patients. Psychiatr Res 20: 33–46

Suddath RL, Christison GW, Torrey EF, Casanova MF, Weinberger DR (1990) Anatomical abnormalities in the brains of monozygotic twins discordant for schizophrenia. N Engl J Med 322: 789–794

Weinberger DR (1987) Implications of normal brain development for the pathogenesis of schizophrenia. Arch Gen Psychiatry 44: 660–669

Weinberger DR, Torrey EF, Neophytides AN, Wyatt RJ (1979) Structural abnormalities in the cerebral cortex of chronic schizophrenic patients. Arch Gen Psychiatry 36: 935–939

Weinberger DR, DeLisi LE, Perman GP, Targum S, Wyatt RJ (1982) Computed tomography in schizophreniform disorder and other acute psychiatric disorders. Arch Gen Psychiatry 39: 778–783

Weinberger DR, Berman KF, Zec RF (1986) Physiologic dysfunction of dorsolateral prefrontal cortex in schizophrenia. I. Regional cerebral blood flow evidence. Arch Gen Psychiatry 43: 114–124

Williams AO, Reveley MA, Kolakowska T, Ardern M, Mandelbrote BM (1985) Schizophrenia with good and poor outcome. II. Cerebral ventricular size and its clinical significance. Br J Psychiatry 146: 239–246

Young AH, Blackwood DH, Roxborough H, McQueen JK, Martin MJ, Kean D (1991) A magnetic resonance imaging study of schizophrenia: brain structure and clinical symptoms. Br J Psychiatry 158: 158–164

Author's address: Prof. Dr. R. E. Gur, Department of Psychiatry, University of Pennsylvania, 10 Gates Building, 36th and Spruce Streets, Philadelphia, PA 19104, U.S.A.

J Neural Transm (1992) [Suppl] 36: 23–40

Molecular insights into schizophrenia

J. W. Pettegrew[1] and **N. J. Minshew**[2]

[1]Neurophysics Laboratory and [2]Autism and Social Disabilities Program, Center for Membrane Studies, Laboratory of Neurophysics, Department of Psychiatry, Western Psychiatric Institute and Clinic, School of Medicine, University of Pittsburgh, Pittsburgh, PA, U.S.A.

Summary. A number of studies have demonstrated alterations in the structure and function of the frontal cortex in some schizophrenic patients. The possible etiology and pathogenesis of these abnormalities are unknown, but genetic and developmental causes are frequently mentioned. Recent in vivo ^{31}P NMR studies of the dorsal prefrontal cortex have been conducted in eleven neuroleptic naive, first episode schizophrenic patients and compared with normal controls of comparable age, educational level and parental educational level. The findings in the schizophrenic patients are different from those of normal IQ adult autistic patients of comparable age and Alzheimer's patients but similar to normal elderly controls. These studies show decreased frontal lobe utilization of adenosine triphosphate in the schizophrenic patients which suggests a hypoactive dorsal prefrontal cortex. In addition, indices of membrane phospholipid metabolism are altered in the schizophrenic patients. However, the findings in the schizophrenic patients are quite similar to those observed in normal elderly controls and to those that normally occur to a lesser degree during adolescence. The phospholipid alterations observed in the schizophrenic patients are compatible with either premature aging or altered timing and exaggeration of the regressive events which occur during normal brain development. The changes in high-energy phosphate metabolism observed in the schizophrenic patients may prove to be state dependent, but the changes in membrane phospholipid metabolism could be related to molecular changes that precede the onset of clinical symptoms and brain structural changes in schizophrenia. These findings suggest new avenues of thinking about the pathogenesis and treatment of schizophrenia.

Introduction

Evidence for impaired frontal lobe function in schizophrenia

Several lines of evidence indicate impaired frontal lobe function ("hypofrontality") in some schizophrenic patients (Buchsbaum et al., 1982;

Weinberger et al., 1986). Cerebral blood flow studies using ^{133}Xenon demonstrate a reduced ratio of frontal lobe to whole surface blood flow in schizophrenic patients under resting conditions (Ingvar and Franzen, 1974) and during cortical activation with the Wisconsin Card Sort (Berman et al., 1988). Reduced ratio of frontal to occipital metabolic rate also is observed in some schizophrenics (Buchsbaum, 1987). These studies suggest that further metabolic studies of the frontal lobe might enhance our understanding of the pathophysiology of schizophrenia. Since normal brain function is dependent on normal high energy phosphate and membrane phospholipid metabolism, either or both of these aspects of brain metabolism might be altered in schizophrenia.

Evidence for membrane alterations in schizophrenia

Other evidence links possible membrane related abnormalities to the pathophysiology of major psychoses. Several studies examining erythrocyte membrane phospholipids have shown phosphatidylcholine (PtdC) to be reduced in some patients with schizophrenia (Stevens, 1972; Henn, 1980; Hitzemann et al., 1984). Increases in phosphatidylserine (PtdS) and smaller decreases in phosphatidylethanolamine (PtdE) also have been noted, but less consistently replicated (Rotrosen and Wolkin, 1987). Such abnormalities appear to be independent of previous drug treatment (Stevens, 1972). Phospholipase A_2 activity, which alters phospholipid fatty acid composition through the deacylation-reacylation cycle, has been found to be increased in some schizophrenics (Gattaz, 1987). In addition, some investigators (Kaiya et al., 1984) have described abnormalities in the phosphatidylinositol (PtdI) cycle in a subgroup of schizophrenic patients. There are, therefore, reasons to suspect alterations in brain membrane phospholipid metabolism in some patients with schizophrenia. However, to date, these parameters have only been examined in peripheral tissues. Support for drawing analogies between peripheral blood elements and brain tissue is derived from observed membrane abnormalities in peripheral cells in Huntington's chorea, myotonic dystrophy, affective illness and Alzheimer's disease (Butterfield et al., 1977, 1978, 1985; Pettegrew et al., 1979a,b,c, 1981, 1982, 1983a; Butterfield and Markesbery, 1980; Markesbery et al., 1980; Blass et al., 1985; Sherman et al., 1986; Diamond et al., 1983; Zubenko et al., 1987; Miller et al., 1989).

Results and discussion: ^{31}P NMR assessment of brain high-energy phosphate and membrane phospholipid metabolism

Significance of in vitro ^{31}P NMR studies

Phosphorus-31 nuclear magnetic resonance spectroscopy (^{31}P NMR) has proven to be a powerful analytical method for investigating phosphorus metabolism in neural and extra-neural tissues (Barany and Glonek, 1984). Studies on neural tissues have utilized tissue extracts (Pettegrew et al., 1979c; Glonek et al., 1982), in vitro brain slices (Cohen et al., 1984) and in vivo studies of animals and humans (Chance et al., 1978; Ackerman et al., 1980; Cady et al., 1982; Petroff et al., 1984; Maris et al., 1985).

The in vitro analytical studies provide chemical conditions that are more favorable for ^{31}P NMR analysis than those occurring in the living brain and, therefore, achieve greater sensitivity and resolution. The enhanced sensitivity and resolution of in vitro extract studies allows the characterization and quantitation of many different phosphorus-containing compounds. Results from the in vitro studies are very important in order to properly interpret in vivo findings in which the resolution and peak dispersion are reduced. Data from previous ^{31}P NMR studies demonstrated a remarkable correlation with those obtained from more classical assay procedures, and, in addition, revealed previously-uncharacterized metabolites and unrecognized metabolic relationships (Glonek et al., 1982; Cohen et al., 1984). The details of the harvesting of brain tissue, perchloric acid extraction and ^{31}P NMR analyses have been previously published (Pettegrew et al., 1987a, 1988b, 1990).

In order to interpret ^{31}P NMR spectra correctly, the identities of the individual resonance signals must be carefully verified through the use of appropriate biochemical and spectroscopic procedures. The importance of this verification was recently demonstrated for a prominent ^{31}P NMR resonance at 3.84 ppm (3.84δ) in mammalian brain which was identified as phosphoethanolamine (Pettegrew et al., 1986). The identification was based on ^{1}H and ^{31}P NMR findings (including pH titrations) at 4.7 and 14.1 Tesla, as well as thin-layer chromatographic analysis. In addition, L-phosphoserine is a significant contributor to the phosphomonester resonance region in mammalian, including human, brain (Pettegrew et al., 1990).

A representative high-resolution ^{31}P NMR spectrum of a brain perchloric acid extract is shown in Fig. 1. The easily identifiable resonances and their chemical shifts (δ) include the phosphomonoesters (PME): α-glycerol phosphate (α-GP, 4.29δ), phosphoethanolamine (PE, 3.84δ), phosphoserine (PS, 3,89δ) and phosphocholine (PC, 3.33δ); inorganic orthophosphate (Pi, 2.63δ); the phosphodiesters (PDE): glycerol 3-phosphoethanolamine (GPE, 0.81δ) and glycerol 3-phosphocholine (GPC, −0.13δ); phosphocreatine (PCr, −3.12δ); the nucleotide triphosphates (especially ATP; γ −5.80δ, α −10.92δ, β −21.4δ); the nucleotide diphosphates (esp. ADP; β −6.11δ, α −10.61δ); dinucleotides such as nicotinamide adenine dinucleotide (NAD) (−11.37δ) and a complex resonance band centered around −12.89δ which is composed of nucleoside diphospho- derivatives such as uridine diphosphosugars and cytidine diphospho- derivatives such as cytidine diphosphocholine and cytidine diphosphoethanolamine.

The PME phosphocholine, phosphoethanolamine and L-phosphoserine are normally found in abundance in mammalian brain, including human, and are important constituents in the metabolism of their respective phospholipids. The brain concentrations (mM) are: phospohoethanolamine

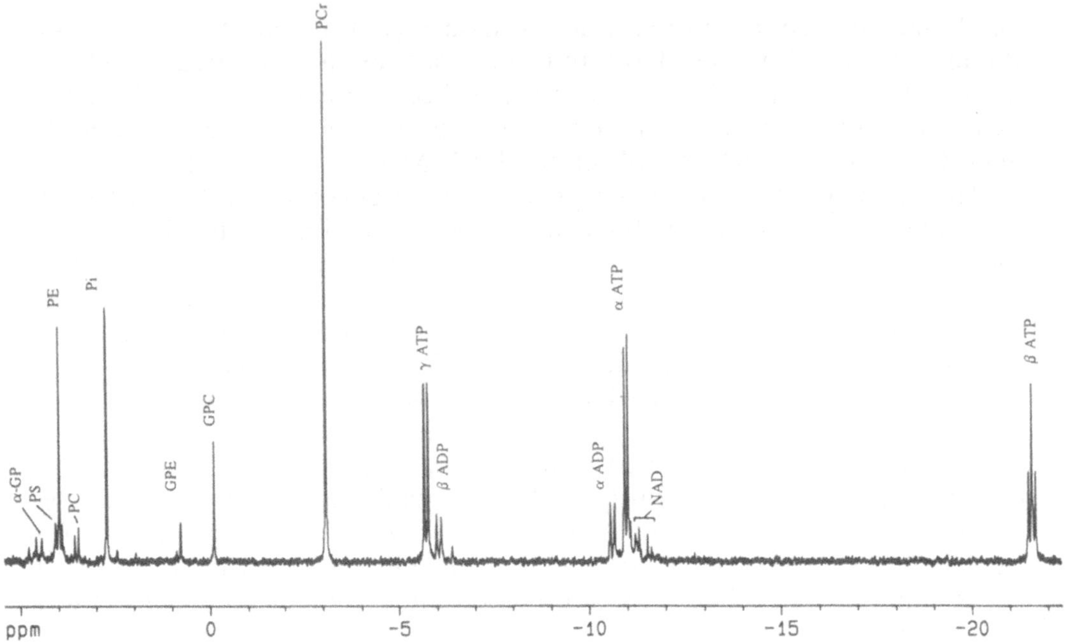

Fig. 1

≃ 1.5, phosphocholine ≃ 0.5 and L-phosphoserine ≃ 0.3. The levels of the PME are increased approximately three-fold in the developing brain during neuritic proliferation (Pettegrew et al., 1990). Similar high levels of PME have been demonstrated in newborn human brain by ^{31}P NMR (Cady et al., 1983). In addition, inositol 1,4,5 triphosphate, Ins(1,4,5)P_3, is a PME and is an important second messenger involved in the regulation of intracellular calcium levels (Berridge and Irving, 1984). Ins(1,4,5)P_3 is derived from PtdI(4,5)P_2 which comprises approximately 1–10% of the total PtdI (Majerus et al., 1986). Since the PtdI concentration in mammalian brain is approximately 1.9 mM (McIlwain and Bachelard, 1985), the normal concentration of PtdI(4,5)P_2 is approximately 0.01–0.1 mM. Therefore, the levels of Ins(1,4,5)P_3 normally found in brain tissue would be below the limits of detection by in vivo ^{31}P NMR. The PME are produced by: (1) phosphorylation of their respective bases by kinases; (2) phospholipase C cleavage of their respective phospholipids or (3) phosphodiesterase cleavage of their respective PDE such as glycerol 3-phosphocholine and glycerol 3-phosphoethanolamine. The PME are broken down by phosphatases or phospholipase D to release Pi and the constituent base.

The PDE glycerol 3-phosphocholine and glycerol 3-phosphoethanolamine also are in high abundance in human brain. The PDE are products of phospholipase A_1 + A_2 activity and are converted to their respective PME by PDE phosphodiesterase activity. In spite of their high abundance in brain, the physiological function of the PDE remains unknown.

In order to understand the meaning and significance of change in the levels of the above metabolites in disease states, a detailed knowledge of the effects of brain development maturation and aging on these metabolites is needed.

Periods of brain development

Brain maturation and development can be arbitrarily divided into four periods (McIlwain and Bachelard, 1985). Period I is the period of cellular division and extends up to birth in the rat. The rat brain at birth has approximately 15% of its adult weight and has no recordable electrical activity. Period II in the rat extends from birth to ten days of age. In Period II, there is growth in the size of individual cells, rapid outgrowth of axons and the rapid development of dendritic connections. During this period there is a rapid increase in brain lipid content which peaks at approximately ten days. During Period III (ten to 20 days of age in the rat) synaptic densities and the number of nerve terminals rapidly increase and electroencephalographic activity (EEG) develops. During this period there are increases in cell volume, and the associated K^+ space with decreases in the extracellular Na^+ and Cl^- spaces. The rates of glucose uptake, glycolysis and oxidative phosphorylation all steadily increase from birth to reach their adult levels during Period III. Naturally occurring cell death has been demonstrated to occur during the period of synaptogenesis which would correspond to Periods II and III in the rat (Pittman and Oppenheim, 1979; Cowan et al., 1984; Oppenheim, 1985; Clarke, 1985). Adenosine triphosphatase activity, including activity requiring added Na^+ and K^+, increases during Periods II and III. Period IV in the rat extends from 20 days and is associated with active myelination but little further brain growth. During this period creatine phosphokinase activity reaches maximal levels. Senescence in rats is thought to develop after 12 months of age and certainly by 24 months of age.

Animal studies of brain development and aging

There are striking effects of animal development and aging on the brain levels of the α-GP, PE, PC, GPE, GPC, PCr and Pi (Pettegrew et al., 1990). The PME (α-GP, PE and PC) are high in the immature brain and rapidly decrease from the newborn period to three months of age ($p = 0.0001$). The levels of α-GP and PC then remain relatively constant until 12 months of age but the levels of PE increase slightly ($0.01 < p < 0.05$). From 12 to 24 months, the levels of PC decrease ($0.0001 < p < 0.001$). In contrast, the PDE (GPE and GPC) are very low or undetectable in the newborn period and then rapidly rise up to three months of age ($p = 0.0001$) after which the levels rise at a slower rate up to 24 months of age.

The PME/PDE ratio, which is an estimate of membrane phospholipid anabolic/catabolic activity (Pettegrew et al., 1987a), is high in the newborn period, rapidly decreases up to three months of age ($p = 0.0001$) and then remains relatively constant up to 12 months of age. After 12 months of age the PME/PDE ratio decreases slightly ($0.01 < p < 0.05$) suggesting that membrane phospholipid breakdown is proceeding slightly faster than membrane phospholipid synthesis.

Possible enzymatic causes for the high PME levels during brain development are: 1) increased choline, ethanolamine or serine kinase activity, 2) increased phospholipase C activity, 3) increased glycerophosphodiesterase activity, 4) decreased CDP-choline phosphotransferase activity, 5) decreased phosphomonoester phosphatase activity or 6) decreased phospholipase D activity. Possible enzymatic causes for the low PDE levels during brain development are: 1) decreased phospholipase A_1 and A_2 activity or decreased glycerophosphodiesterase activity.

These ^{31}P NMR results tentatively suggest very high phospholipid anabolic activity without appreciable catabolic activity from 12 h to ten days of age. This would coincide with an increase in the size of neuronal cells, the rapid outgrowth of axons and the rapid development of dendritic connections reported to occur in Period II. After ten days of age, there is a rather rapid increase in PDE which are catabolic breakdown products of membrane phospholipids. The PME levels start to decrease at ten days of age, but the PME/PDE ratio remains high (>10). The increasing PDE observed from ten to 30 days of age could represent the metabolic correlates of naturally occurring cell death and the remodelling of neuritic connections. Cell death has been demonstrated to occur naturally during neural development and, in particular, occurs at about the time of rapid synaptogenesis which is from birth to 20 days of age in the rat (Pittman and Oppenheim, 1979; Cowan et al., 1984; Oppenheim, 1985; Clarke, 1985). From one to three months of age the PME continue to decrease and the PDE continue to increase, although not as rapidly as before. The PME/PDE ratio falls from ten to two indicating that the rate of phospholipid synthesis is decreasing. The changes in PME and PDE occurring from one to three months of age could reflect active myelination which is known to occur during this time in the rat. After three months of age, the turnover rate for phospholipids (PME/PDE) remains relatively constant until 12 months of age. From 12 to 24 months of age, the PME/PDE ratio decreases further ($0.01 < p < 0.05$) suggesting that membrane catabolism is proceeding slightly faster than anabolism.

PCr is the most metabolically labile of the brain high-energy phosphates. Levels of PCr are quite low in the newborn period and appear to decrease even further up to five days of age. After five days of age, the PCr levels rapidly increase up to one month of age, with less rapid increases between one and three months of age (12 h versus three month $p = 0.0001$). After three months of age, the PCr levels undergo less dramatic but steady increases up to 24 months of age. In contrast, the levels of Pi are relatively high in the newborn, decrease up to five days of age followed by increases up to ten days of age. After ten days of age, the Pi levels drop rapidly until three months of age ($0.01 < p < 0.05$). From three months of age until 12 months of age, the Pi levels appear to increase slightly with more rapid increases from 12 to 24 months of age.

The PCr/Pi ratio provides a convenient measure of the energy status of the brain as it is the ratio of the most labile form of high-energy phosphate (PCr) to the ultimate breakdown product of all high-energy phosphates (Pi). The PCr/Pi ratio is quite low in the newborn period until five days of age after which time the PCr/Pi ratio rapidly increases up to three months of age ($p = 0.0001$). After three months of age, the PCr/Pi ratio remains relatively constant up to 24 months of age. The rapid increase in the PCr/Pi ratio up to one month of age correlates with the development of the glycolytic and oxidative pathways, increasing Na^+-K^+ ATPase activity and the onset of electroencephalographic activity. The PCr/Pi ratio appears to increase slightly after 12 months of age suggesting decreased utilization of PCr. The brain levels of ATP do not undergo similar changes; there are no significant differences in brain ATP content comparing animals of 12 hours versus three months, three months versus 12 months, or 12 months versus 24 months of age.

In vivo ^{31}P NMR studies

Brain chemistry also can be assessed in vivo using ^{31}P NMR spectroscopy. ^{31}P has a nuclear magnetic moment and is present in 100% natural abundance so no isotope has to be given to the subject. ^{31}P NMR spectroscopy provides a direct in vivo assessment of the brain membrane phospholipid and high energy phosphate metabolism. The PME and PDE resonances in mammalian brain originate predominantly from the precursors and breakdown products respectively of membrane phospholipids (Pettegrew et al., 1986, 1987b). The PCr, ATP, ADP and Pi levels reflect the state of energy metabolism (Pettegrew et al., 1987b). In addition, the intracellular pH can be directly assessed (Petroff et al., 1985; Pettegrew et al., 1988a). As NMR spectroscopy is safe and non-invasive, repeated measurements can be carried out in the same individual over a period of time making longitudinal studies possible.

In vivo ^{31}P NMR spectra on human subjects have been obtained using a General Electric Signa System with the spectroscopy research accessory. The field strength is 1.5 Tesla, yielding a proton frequency of 63.970 MHz, and a phosphorus frequency of 25.895 MHz. The subject is placed supine on the NMR magnet transport table, and the head is positioned inside a plexiglass head holder containing a support bracket for securing a surface coil. The surface coil is positioned over the head so as to sample signals from the dorsal prefrontal cortex.

All subjects have routine T_1 weighted ^1H MRI scans in the sagittal, coronal and axial planes immediately preceding or following the ^{31}P NMR spectroscopy studies. ^1H images, which are used to identify the location and volume of brain sampled by spectroscopy as well as the ^{31}P NMR spectra, are acquired with the same probe containing an 8 inch ^{31}P surface coil and a 3 inch surface coil dual-tuned at both the ^{31}P and ^{1}H frequencies. ^{1}H images for spectral localization are obtained by transmitting with a Helmholtz body coil, and receiving with the 3 inch surface coil. The localized ^{1}H images and the ^{31}P NMR spectra are from approximately 15–20 cm^3 of the dorsal prefrontal cortex. The B_o field is then shimmed on the H_2O ^{1}H signal to a line width of approximately 0.1 ppm.

^{31}P NMR spectra are obtained using a B_1 field gradient and a pulse-acquire sequence. A pulse width of 800 μsec produces a spectrum with PCr/Pi and PME/PDE ratios typical of mammalian brain. The importance of these considerations have previously been pointed out (Pettegrew et al., 1983b). Other acquisition parameters are as follows: time to pulse repetition (TR) = 2 sec, sweep width = 2 kHz, receiver filter = 1 kHz, number of complex data points = 2,048. The integrated spectral areas are the same for TR values of 2, 4 and 6 sec. Therefore, a TR of 2 seconds is used allowing a greater signal to noise for a fixed total acquisition time without signal saturation.

All spectra are processed on a Nicolet data station with 5 Hz exponential multiplication, first and second order phase correction to bring all peaks into absorption mode, and baseline correction by means of a series of linear tilts between known baseline points. Integrated areas are then calculated using the GENCAP program which fits the spectrum with a series of Lorentzian lines. Known doublets, such as the ionized ends region, are fitted with two Lorentzians, and known triplets, such as the middles region, are fitted with three Lorentzians. The PME and PDE peaks are fitted with one, two, or three Lorentzian lines to obtain the most accurate fit. For each spectrum, the integrated areas of the PME, Pi, PDE, PCr, ionized ends (γATP + βADP), esterified ends (αATP + αADP) and middles (βATP) regions are determined. From these integrated areas, the mole % of PME, Pi, PDE, PCr, ATP and ADP are calculated as the integrated area under the resonance peak divided by the total spectral integrated area. The intracellular pH is determined by the chemical shift difference between the PCr and Pi peaks (Petroff et al., 1985) or between the γ and α peaks of ATP (Pettegrew et al., 1988a). Our in vivo results compare quite favorably with previously published in vitro results (Glonek et al., 1982; Cohen et al., 1984; Pettegrew et al., 1987b; Pettegrew, 1990) verifying the validity of the in vivo method used in the present study. Test-retest and inter-rater reliability results give standard deviations of less than 5% of the mean for each parameter (N = 10).

Comparison of in vitro and in vivo ^{31}P NMR findings

The PME, Pi, PDE and PCr resonances of brain are not completely resolved in the unprocessed in vivo ^{31}P NMR spectrum obtained at 1.5 Tesla.

This is probably due to a combination of at least four factors: (1) The NMR natural line widths for these in vivo brain chemical species are probably broader than for the same chemical species in solution indicating longer correlation times (τ_c) for the species in vivo; (2) Some of these chemical species could be in chemical exchange with divalent cations such as Ca^{2+} and Mg^{2+}. If the chemical exchange rate is intermediate with respect to the observe frequency (ω_0), this could lead to line broadening (Pettegrew et al., 1988a). The PME, PDE and Pi have been shown to chelate di- and trivalent cations in solution (Panchalingam et al., 1987) so this possibility also could exist in vivo; (3) There is decreased chemical shift dispersion at 1.5 Tesla compared to B_o fields of 4.7 Tesla or higher; and (4) The B_o field homogeneity across the human head is not as good as that attainable across smaller samples (5–20 mm diameters) with better dielectric homogeneity.

Normal brain development and aging

In order to properly interpret the findings in the schizophrenics, it is important to know how high-energy phosphate and membrane phospholipid metabolism change in normal human brain development and aging. In vivo ^{31}P NMR of the dorsal prefrontal cortex was performed on 29 normal volunteers ages 12 to 81 years. All subjects had normal medical and neurological examinations, no subject meet DSM III-R criteria for any psychiatric disorder and no subject had a positive family history for neuropsychiatric disorders. There was a significant decrease in the levels of PME with age ($p = 0.01$; $r = 0.52$) and a significant increase in the levels of PDE ($p = 0.01$; $r = 0.52$). There were no age-related changes in the levels of PCr, Pi or ATP (Panchalingam et al., 1990).

The studies in normal human volunteers show similar effects of brain development and aging as demonstrated in the Fischer 344 rat. The PME increase between the ages of ten to 20 years remain relatively stable up to 40–50 years and then slowly decline after 50 years of age. The PDE decrease between the ages of ten to 20 years remain relatively stable up to 40–50 years and then slowly increase over the age of 50 years. The changes in brain levels of PME and PDE in individuals between ten to 20 years of age probably reflect increasing membrane anabolic and decreasing membrane catabolic activity between these ages. The decreasing PME and increasing PDE that occurs over the age of 50 presumably reflects the loss of dendritic spines and processes that occur with aging.

Schizophrenics and controls

Brain high energy phosphate and membrane phospholipid metabolism were investigated in the dorsal prefrontal cortex of neuroleptic naive, first episode schizophrenic patients and age, sex, race, education and parental education matched healthy controls under resting conditions (Pettegrew et al., 1991). Neuroleptic naive patients were chosen because of the known lasting effects of neuroleptics on membrane phospholipids (Essali et al., 1989). Patients at an early stage of the illness were chosen to exclude the possible effects of length of illness and hospitalization. Eleven schizophrenic patients (7 males, 4 females; 24.4 ± 1.4 yrs.; mean ±SEM; 7 white, 4 black; 11.7 ± 0.8 yrs. education) and ten matched healthy controls (6 males, 4 females; 24.1 ± 1.8 yrs.; 7 white, 2 black, 1 oriental; 12.7 ± 0.3 yrs. education) were studied. Patients were assessed by the Schedule for Affective Disorders and Schizophrenia (SADS) (Endicott and Spitzer, 1978) and met both DSM-III-R (1987) and Research Diagnostic Criteria (RDC) for schizophrenia (Spitzer et al., 1990). The DSM-III-R subclassification for the patients was 6 undifferentiated and 5 paranoid. None were previously treated with neuroleptics as reported by the patients' families or the referring physicians. The mean illness duration was 19.1 ± 5.9 months. None had a history of persistent substance abuse or met DSM-III-R criteria for it. Physical examination did not reveal abnormalities in any patient. The patients were assessed further using the Brief Psychiatric Rating Scale (BPRS) (Overall and Gorham, 1962). All the patients have been followed for at least six months after entry into the study to ensure diagnostic stability. The control subjects were selected from among students and hospital staff. Psychiatric, medical, and family history information were obtained using a structured form. All patients and controls signed an informed consent. There are no significant group differences between the patients and controls for age, sex, race, education or parental education. The MRI results are normal in the schizophrenics and controls as reported by experienced neuroradiologists. The following metabolic alterations are observed in the dorsal prefrontal cortex of the schizophrenic patients: (1) decreased levels of PME (schizophrenics = 15.5 ± 0.9; controls = 20.3 ± 0.9; p = 0.002); (2) increased levels of PDE (schizophrenics = 36.4 ± 1.5; controls = 32.2 ± 0.6; p = 0.02); (3) increased levels of ATP (schizophrenics = 9.6 ± 0.5; controls = 7.4 ± 0.4; p = 0.02) and (4) decreased levels of P_i (schizophrenics = 7.9 ± 0.5; controls = 9.9 ± 0.5; p = 0.01). There were no group differences in the levels of PCr, ADP or intracellular pH.

Possible causes for decreased observable levels of PME in schizophrenia brain are: (1) decreased kinase activity; (2) decreased phospholipase C activity; (3) decreased PDE phosphodiesterase activity; (4) increased phospholipase D activity; (5) increased chemical exchange with divalent cations leading to decreased NMR observability or (6) increased τ_c with resulting decreased NMR observability.

Possible causes for increased PDE in schizophrenic brain include decreased PDE phosphodiesterase activity and increased phospholipase A_1 + A_2 activity. Gattaz (1987) has reported increased activity of phospholipase A_2 in schizophrenia. Other possible causes of the increased observable PDE are decreased τ_c with increased NMR observability or decreased chemical exchange with divalent cations resulting in increased NMR observability. Decreased PDE phosphodiesterase activity could account for both the decreased PME and increased PDE levels observed in the schizophrenic patients studied.

The present findings of decreased PME and increased PDE levels also could suggest decreased synthesis and increased breakdown of membrane phospholipids in schizophrenia. Similar results are observed in normal aging in animals by in vitro ^{31}P NMR (Cady et al., 1983) and humans by in vivo ^{31}P NMR spectroscopy (Panchalingam et al., 1990). Demisch et al. (1987) have reported a significantly decreased incorporation of 14-C labelled arachidonic acid in PtdC, PtdE and PtdI in patients with schizoaffective and schizophreniform disorders which could suggest decreased membrane phospholipid synthesis. As previously mentioned, decreased levels of membrane PtdC have been observed in erythrocytes in some schizophrenics (Stevens, 1972; Henn, 1980; Hitzemann et al., 1984). All the schizophrenic patients in the present study had normal MRI findings (as reported by neuroradiologists) at the time of their ^{31}P NMR studies. It is possible, therefore, that the PME and PDE alterations could represent metabolic alterations which antedate the onset of anatomical changes.

The increased ATP and decreased P_i observed in the present study suggests decreased ATP utilization with decreased production of P_i. These findings are consistent with hypoactivity of the dorsal prefrontal cortex as suggested by decreased cerebral blood flow and decreased uptake of 2-deoxyglucose (Ingvar and Franzen, 1974; Buchsbaum et al., 1982; Weinberger et al., 1986; Berman et al., 1988). Both cerebral blood flow and 2-deoxyglucose uptake are indirect measures of cerebral energy metabolism. The uptake of glucose and presumably 2-deoxyglucose are dependent upon normal membrane properties (Carruthers et al., 1990), and the decreased uptake of 2-deoxyglucose observed in some schizophrenics also could reflect altered membrane phospholipids.

Schizophrenics and autistics

Normal IQ, adult autistic patients, who are similar in age to the schizophrenic patients and share some of the negative symptoms, were studied by in vivo ^{31}P NMR spectroscopy using the identical protocol used for the schizophrenic patients. Adult autistic patients, who are similar in age to the schizophrenic patients in this study and share the negative symptoms of schizophrenia, do not have altered levels of brain PME or PDE by in vivo ^{31}P NMR (Pettegrew et al., 1989; Minshew et al., 1990). However, the autistic patients have decreased levels of ATP which is the opposite of that observed in the schizophrenic patients. The decreased levels of ATP suggests increased ATP utilization in the autistics. This is consistent with a PET study of 12 young adult autistic men which reported an increase in metabolic rate for glucose in frontal, parietal, temporal and occipital cortex, hippocampus, thalamus and basal ganglia (Rumsey et al., 1985).

Schizophrenia, normal aging and Alzheimer's disease

It is of interest that the changes in PME and PDE observed in schizophrenics are distinct and in the opposite direction from those observed with Alzheimer's disease which is associated with increases in brain PME early in the course of the disease. Elevated brain levels of PDE occur later in the course of Alzheimer's disease and probably reflect the onset of membrane degenerative changes (Brown et al., 1989; Pettegrew et al., 1984, 1987b, 1988b,c, 1990). As previously mentioned, the changes observed in schizophrenia are quite similar to those observed in normal aging in which the brain levels of PME are decreased and the PDE are increased (Pettegrew et al., 1987b). From this perspective, schizophrenia could be considered a form of premature brain aging, perhaps involving only certain neural systems. This formulation would be in keeping with the concept of dementia praecox put forth many years ago (Kraeplin, 1971; Morel, 1860). All the schizophrenic patients in the present study had normal MRI findings, as reported by neuroradiologists, at the time of their ^{31}P NMR studies. It is possible, therefore, that the PME and PDE alterations could represent metabolic alterations which antedate the onset of anatomical changes. This could provide a possible metabolic explanation for the development of cerebral atrophy observed in some schizophrenics (Andreasen et al., 1986; Andreasen, 1988), including the affected twin of monozygotic twins discordant for schizophrenia (Suddath et al., 1990).

Schizophrenia and programmed synaptic pruning

The findings of decreased PME and increased PDE in first-episode, drug naive schizophrenic patients also could be due to abnormal brain development as suggested by Feinberg (Feinberg, 1982; Weinberger, 1987). There is now substantial evidence for programmed neuronal cell loss, loss of callosal axons and synaptic elimination in normal brain development in animals and humans (Huttenlocher et al., 1982; Purves, 1980; Huttenlocher, 1979; Rakic, 1983; Cowan, 1984). The present in vivo ^{31}P NMR findings indicate that the cell bodies, processes and terminals in the sampled area of the dorsal prefrontal cortex of the schizophrenic patients contain depressed levels of PMEs and elevated levels of PDEs as compared to the control subjects. In vivo ^{31}P NMR studies of normal volunteers between the ages of 12 and 85 years also reveal a transient decrease in the levels of PME and an increase in the levels of PDE during adolescence (10–12 years) (Panchalingam et al., 1990). However, the decreased PME and increased PDE observed in the schizophrenic patients appear to be more pronounced than in the normal volunteers of comparable age. This suggests decreased synthesis and enhanced breakdown of membranes occurring in the schizophrenic patients. This could be due to altered timing or abnormal enhancement of normal programmed synaptic pruning during adolescence as initially suggested by Feinberg (Feinberg, 1982) and demonstrated to occur in the prefrontal cortex of humans during adolescence by Huttenlocher (Purves, 1980). An exaggeration of these regressive synaptic events involving apical and basal dendrites could produce brain structural changes similar to those observed in schizophrenia in which the predominant reduction is in grey but not white matter. An exaggeration of normal neuronal cell death should produce decreased numbers of neurons and their projection axons and, therefore, reduce the volume of both grey and white matter.

Schizophrenia and prefrontal glutamatergic pathways

The descending prefrontal glutamatergic pathway is a major input to both the striatum and the limbic cortex. Exaggerated synaptic pruning of prefrontal-limbic glutamatergic axonal terminals could produce a reduced neuropil volume resulting in what might appear as an enhanced clustering of cells as observed for pre-α-cells in the parahippocampal gyrus (Falkai, 1988). Excessive pruning of prefrontal-striatal glutamatergic axonal terminals ending on striatal dopaminergic terminals could decrease the tonic release of dopamine from the dopaminergic terminals resulting in secondary up-regulation of the postsynaptic dopaminergic receptors. A number of postmortem studies have demonstrated increased numbers of D_2 receptors in the caudate and putamen of patients with schizophrenia (Seeman et al., 1984; Jaskiw, 1988), but a recent study suggests this is secondary to neuroleptic treatment within three months of death (Kornhuber et al., 1989). Two in vivo positron emission tomography studies of brain D_2 receptors in schizophrenic patients have appeared. One study reported increased numbers of D_2 receptors in the striatum of drug-naive schizophrenic patients (Wong et al., 1986), but the other study could not replicate this finding (Farde et al., 1987). The phasic release of dopamine is thought to be responsive to stress. With up-regulation of postsynaptic dopaminergic receptors, stress could result in exaggerated mesolimbic activity resulting in positive symptoms. Neuroleptics which block postsynaptic D_2 receptors could modulate the mesolimbic activity back towards normal. This formulation has been suggested recently by Grace (1991).

Potential therapeutic effect of neurotrophic agents

Premature aging of the prefrontal cortex or an exaggeration or altered timing of the normal synaptic pruning of the prefrontal cortex, which occurs in adolescence, could potentially produce the PME and PDE findings observed in the schizophrenic patients. Should decreased brain PME and increased PDE be found in a much larger sample of schizophrenic patients, then consideration should be given to the possible use of brain trophic factors early in the course of schizophrenia. The goal would be to identify by ^{31}P NMR those individuals who demonstrate the PME and PDE changes early in the course of the illness or even better prior to the onset of symptoms. These individuals would then be candidates for neurotrophic therapy to slow, stop or, hopefully, reverse the membrane alterations. This is obviously highly speculative, and even if this approach should prove beneficial, it is still far into the future. However, by starting to plan for the future now, perhaps we can make the future less far off.

The ability of ^{31}P NMR to monitor high energy phosphates as well as membrane phospholipid metabolites could provide new insights not previously available with other techniques. ^{31}P NMR spectroscopy also appears to be a practicable procedure for longitudinal studies which can potentially provide new information on the possible evolution of regional biochemical changes in schizophrenia.

Acknowledgements

Ms. M. Young is gratefully acknowledged for typing the manuscript and for her helpful editorial assistance.

References

Ackerman JH, Grove TH, Wong GG, Gadian DG, Radda GK (1980) Mapping of metabolites in whole animals by 31P nuclear magnetic resonance spectroscopy. Nature 283: 167–170

Andreasen NC (1988) Brain imaging: applications in psychiatry. Science 239: 1381–1388

Andreasen N, Nasrallah HA, Dunn V, Olson SC, Grove WM, Ehrhardt JC, Coffman JA, Crossett JHW (1986) Structural abnormalities the frontal system in schizophrenia. Arch Gen Psychiatry 43: 136–144

Barany M, Glonek T (1984) Identification of diseased states by phosphorus-31 NMR. In: Gorenstein DG (ed) Phosphorus-31 NMR, principles and applications. Academic Press, New York, pp 511–515

Berman KF, Illowsky BP, Weinberger DR (1988) Physiological dysfunction of dorsolateral prefontal cortex in schizophrenia. Arch Gen Psychiatry 45: 616–622

Berridge MJ, Irvine RF (1984) Inositol trisphosphate, a novel second messenger in cellular signal transduction. Nature 312: 315–321

Blass JP, Hanin I, Barclay L, Kopp U, Reding MJ (1985) Red blood cell abnormalities in Alzheimer's disease. J Am Geriatr Soc 33:401–405

Brown GG, Levine SR, Gorell JM, Pettegrew JW, Gdowski JW, Bueri JA, Helpern JA, Welch KMA (1989) In vivo 31P NMR profiles of Alzheimer's disease and multiple subcortical infarct dementia. Neurology 39: 1423–1427

Buchsbaum MS (1987) Positron emission tomography in schizophrenia. In: Meltzer HY (ed) Psychopharmacology, the third generation of progress. Raven Press, New York, pp 783–792

Buchsbaum MS, Ingvar DH, Kessler R, Waters RN, Cappelette J, van Kammen DP, King C, Johnson JL, Manning RB, Flynn RW, Mann LS, Bunney WE, Sokoloff L (1982) Cerebral glucography with positron tomography. Arch Gen Psychiatry 39: 251–259

Butterfield DA, Markesbery WR (1980) Specificity of biophysical and biochemical alterations in erythrocyte membranes with neurological disorders. J Neurol Sci 97: 261–271
Butterfield DA, Oeswein JW, Markesbery WR (1977) Electron spin resonance study of membrane protein alterations in erythrocytes in Huntington's disease (letter). Nature 267: 453–455
Butterfield DA, Oeswein JW, Prunty ME, Hisle KC, Markesbery WR (1978) Increased sodium plus potassium adenosine triphosphatase activity in erythrocyte membranes in Huntington's disease. Ann Neurol 4: 60–62
Butterfield DA, Nicholas MM, Markesbery WR (1985) Evidence for an increased rate of choline efflux across erythrocyte membranes in Alzheimer's disease. Neurochem Res 10: 909–918
Cady EB, Dawson MJ, Hope PL, Tofts PS, Costello AM, Delpy DT, Reynolds EOR, Wilkie DR (1983) Non-invasive investigation of cerebral metabolism in newborn infants by phosphorus nuclear magnetic resonance spectroscopy. Lancet i: 1059–1062
Carruthers A, Helgerson AL, Herbert DN, Tefft RE Jr, Naderi S, Melchior DL (1989) Effects of calcium, ATP and lipids on human erythrocyte sugar transport. Ann NY Acad Sci 568: 52–67
Chance B, Nakase Y, Bond M, Leigh JS Jr, McDonald G (1978) Detection of ^{31}P nuclear magnetic resonance signals in brain by in vivo and freeze trapped assays. Proc Natl Acad Sci USA 75: 4925–4929
Clarke PG (1985) Neuronal death in the development of the vertebrate nervous system. Trends Neurosci 8: 345–349
Cohen MM, Pettegrew JW, Kopp SJ, Minshew N, Glonek T (1984) P-31 nuclear magnetic resonance analysis of brain: normoxic and anoxic brain slices. Neurochem Res 9: 785–801
Cowan WM, Fawcett JW, OLeary DD, Stanfield BB (1984) Regressive events in neurogenesis. Science 225: 1258–1265
Demisch L, Gerbaldo H, Heinz K, Kirsten R (1987) Transmembranal signalling in schizophrenic and affective disorders: studies on arachidonic acid and phospholipids. Schizophr Res 22: 275–282
Diamond JM, Matsuyama SS, Meier K, Jarvik LF (1983) Elevation of erythrocyte countertransport rates in Alzheimer's dementia (letter). N Engl J Med 309: 1061–1062
DSM-III-R Diagnostic and Statistical Manual of Mental Disorders, 3rd edn (1987) American Psychiatric Association. Workgroup to revise DSM-III, Washington, DC
Endicott J, Spitzer RL (1978) A diagnostic interview: the schedule for affective disorders and schizophrenia. Arch Gen Psychiatry 35: 837–844
Essali MA, Das I, deBelleroche J, Hirsch SR (1989) The platelet polyphosphoinositide system in schizophrenia: pathological and pharmacological implications. Schizophr Res 2: 148
Falkai P, Bogerts B, Rozumek M (1988) Cell loss and volume reduction in the entorhinal cortex of schizophrenics. Eur Arch Psychiatry Neurol Sci 24: 515–521
Farde L, Wiesel FA, Hall H, Halldin C, Stone-Elander S, Sedvall G (1987) No D2 receptor increase in PET study of schizophrenia (letter). Arch Gen Psychiatry 44: 671–672
Feinberg I (1982) Schizophrenia: caused by a fault in programmed synaptic elimination during adolescence? J Psychiatr Res 17 [Suppl] 4: 319–334
Gattaz W (1987) Increased plasma phospholipase A2 activity in schizophrenic patients: reduction in schizophrenia. Biol Psychiatry 22: 421
Glonek T, Kopp SJ, Kot E, Pettegrew JW, Harrison WH, Cohen MM (1982) P-31 nuclear magnetic resonance analysis of brain: the perchloric acid extract spectrum. J Neurochem 39: 1210–1219

Grace AA (1991) Phasic versus tonic dopamine release and the modulation of dopamine system responsivity: a hypothesis for the etiology of schizophrenia. Neuroscience 41:1–24

Henn F (1980) Biological concepts of schizophrenia. In: Baxter C, Melnachuk T (eds) Perspectives in schizophrenia research. Raven Press, New York, pp 209–223

Hitzemann R, Hirschwitz D, Garver D (1984) Membrane abnormalities in the psychoses and affective disorders. J Psychiatr Res 18: 319–326

Huttenlocher PR (1979) Synaptic density in human frontal cortex. Developmental changes and effects of aging. Brain Res 163: 195–205

Huttenlocher PR, deCourten C, Garey LJ, Van Der Loos H (1982) Synaptogenesis in human visual cortex-evidence for synapse elimination during normal development. Neurosci Lett 33: 247–252

Ingvar DH, Franzen G (1974) Abnormalities of cerebral blood flow distribution in patients with chronic schizophrenics. Acta Psychiatr Scand 50: 425–462

Jaskiw G, Kleinman J (1988) Postmortem neurochemistry studies in schizophrenia. In: Schulz SC, Tamminga HK (eds) Schizophrenia: a scientific focus. Oxford University Press, New York, pp 264–273

Kaiya H, Takeuchi K, Namba M, Imai A, Nakashima S, Nozawa Y (1984) Abnormal phosphatidylinositol-cycle of platelet membrane in schizophrenia — a preliminary study. Folia Psychiat Neurol Jpn 38: 437–444

Kornhuber J, Riederer P, Reynolds GP, Beckmann H, Jellinger K, Gabriel E (1989) 3H-spiperone binding sites in post-mortem brains from schizophrenia patients: relationship to neuroleptic drug treatment, abnormal movements, and positive symptoms. J Neural Transm 75: 1–10

Kraepelin E (1919) Dementia praecox and paraphrenia. Churchill Livingstone, Edinburgh

Majerus PW, Connolly TM, Deckmyn H, Ross TS, Bross TE, Ishii H, Bansal VS, Wilson DB (1986) The metabolism of phosphoinositide-derived messenger molecules. Science 234: 1519–1526

Maris JM, Evans AE, McLaughlin AC, D'Angio GJ, Bolinger L, Manos H, Chance B (1985) 31P nuclear magnetic resonance spectroscopic investigation of human neuroblastoma in situ. N Engl J Med 312: 1500–1505

Markesbery WR, Leung PK, Butterfield DA (1980) Spin label and biochemical studies of erythrocyte membranes in Alzheimer's disease. J Neurol Sci 45: 323–330

McIlwain H, Bachelard HS (1985) Biochemistry and the central nervous system, 5th edn. Churchill Livingstone, New York, pp 41–43

Miller BL, Jenden D, Tang C, Read S (1989) Choline and choline-bound phospholipids in aging and Alzheimer's disease (abstract). Neurology 39 [Suppl] 1: 254

Minshew NJ, Pettegrew JW, Panchalingam K (1990) Membrane phospholipid alterations observed in Alzheimer's disease are not present in Down's syndrome (abstract). Biol Psychiatry 27(9A): 123A–124A

Morel BA (1860) Traitement des maladies mentales. Masson, Paris

Oppenheim RW (1985) Naturally occurring cell death during neural development. Trends Neurosci 8: 487–493

Overall JE, Gorham DR (1962) The brief psychiatric rating scale. Psychol Rep 10: 799–812

Panchalingam K, Post JFM, Pettegrew JW (1987) Evidence for increased aluminum binding ligands in Alzheimer's disease (abstract). Neurology 37 [Suppl] 1: 331

Panchalingam K, Pettegrew JW, Strychor S, Tretta M (1990) Effect of normal aging on membrane phospholipid metabolism by 31P in vivo NMR spectroscopy (abstract). Soc Neurosci Abstr 16: 843

Petroff OAC, Prichard JW, Behar KL, Alger JR, den Hollander JA, Shulman RG (1985) Cerebral intracellular pH by 31P nuclear magnetic resonance spectroscopy. Neurology 35: 781–788

Pettegrew JW, Nichols JS, Stewart RM (1979a) Studies of the fluorescence of fibroblasts from Huntington's disease: evidence of a membrane abnormality. N Engl J Med 300: 678
Pettegrew JW, Nichols JS, Stewart RM (1979b) Fluorescence spectroscopy on Huntington's fibroblasts. J Neurochem 33: 905–911
Pettegrew JW, Glonek T, Baskin F, Rosenberg RN (1979c) Phosphorus-31 NMR of neuroblastoma clonal lines: effect of cell confluency state and dibutyryl cyclic AMP. Neurochem Res 4: 795–801
Pettegrew JW, Nichols JS, Stewart RM (1981) Membrane studies in Huntington's disease: steady-state and time-dependent fluorescence spectroscopy of intact lymphocytes. J Neurochem 36: 1966–1976
Pettegrew JW, Nichols JS, Minshew NJ, Rush AJ, Stewart RM (1982) Membrane biophysical studies of lymphocytes and erythrocytes in manic-depressive illness. J Affective Disord 4: 237–247
Pettegrew JW, Minshew NJ, Stewart RM (1983a) Dynamic membrane studies in individuals at risk for Huntington's disease. Life Sci 32: 1207–1212
Pettegrew JW, Minshew NJ, Diehl J, Smith T, Kopp SJ, Glonek T (1983b) Anatomical considerations for interpreting topical P-31 NMR. Lancet ii: 913
Pettegrew JW, Minshew NJ, Cohen MM, Kopp SJ, Glonek T (1984) P-31 NMR changes in Alzheimer's and Huntington's disease brain (abstract). Neurology 34 [Suppl] 1: 281
Pettegrew JW, Kopp SJ, Dadok J, Minshew SJ, Feliksik JM, Glonek T, Cohen MM (1986) Chemical characterization of a prominent phosphomonoester resonance from mammalian brain: 31P and 1H NMR analysis at 4.7 and 14.1 tesla. J Magn Reson 67: 443–450
Pettegrew JW, Kopp SJ, Minshew NJ, Glonek T, Feliksik JM, Tow JP, Cohen MM (1987a) 31P nuclear magnetic resonance studies of phosphoglyceride metabolism in developing and degenerating brain: preliminary observations. J Neuropathol Exp Neurol 46: 419–430
Pettegrew JW, Withers G, Panchalingam K, Post JF (1987b) 31P nuclear magnetic resonance (NMR) spectroscopy of brain in aging and Alzheimer's disease. J Neural Transm [Suppl] 24: 261–268
Pettegrew JW, Withers G, Panchalingam K, Post JFM (1988a) Considerations for brain pH assessment by 31P NMR. Magn Reson Imaging 6: 135–142
Pettegrew JW, Moossy J, Withers G, McKeag D, Panchalingam K (1988b) 31P Nuclear magnetic resonance study of the brain in Alzheimer's disease. J Neuropathol Exp Neurol 47: 235–248
Pettegrew JW, Panchalingam K, Moossy J, Martinez J, Rao G, Boller F (1988c) Correlation of phosphorus-31 magnetic resonance spectroscopy and morphologic findings in Alzheimer's disease. Arch Neurol 45: 1093–1096
Pettegrew JW, Minshew NJ, Payton JB (1989) 31P NMR in normal IQ adult autistics (abstract). Biol Psychiatry 25: 182
Pettegrew JW, Panchalingam K, Withers G, McKeag D, Strychor S (1990) Changes in brain energy and phospholipid metabolism during development and aging in the Fischer 344 rat. J Neuropathol Exp Neurol 49: 237–249
Pettegrew JW, Keshauan MS, Panchalingam K, Strychor S, Kaplan DB,Tretta MG, Allen M (1991) Alterations in brain high-energy phosphate and membrane phospholipid metabolism on first-episode, drug-naive schizophrenics. Arch Gen Psychiatry 48:563–568
Pittman R, Oppenheim RW (1979) Cell death of motoneurons in the chick embryo spinal cord. IV. Evidence that a functional neuromuscular interaction is involved in the regulation of naturally occurring cell death and the stabilization of synapses. J Comp Neurol 187: 425–446
Purves D, Lichtman JW (1980) Elimination of synapses in the developing nervous system. Science 210: 153–157

Rakic P, Riley KP (1983) Overproduction and elimination of retinal axons in the fetal rhesus monkey. Science 219: 1441–1444

Rotrosen J, Wolkin A (1987) Phospholipid and prostaglandin hypothesis in schizophrenia. In: Meltzer HY (ed) Psychopharmacology. The third generation of progress. Raven Press, New York, pp 759–764

Rumsey JM, Duara R, Grady C, Rapoport JL, Margolin RA, Rapoport SI, Cutler NR (1985) Brain metabolism in autism. Resting cerebral glucose utilization rates as measured with positron emission tomography. Arch Gen Psychiatry 42: 448–455

Seeman P, Ulpian C, Bergeron C, Riederer P, Jellinger K, Gabriel E, Reynolds GP, Tourtellotte WW (1984) Bimodal distribution of dopamine receptor densities in brains of schizophrenics. Science 225: 728–731

Sherman KA, Gibson SE, Blass JP (1986) Human red blood cell choline uptake with age and Alzheimer's disease. Neurobiol Aging 7: 205–209

Spitzer RL, Endicott J, Robins E (1990) Research diagnostic criteria (RDC) for a selected group of function disorders, 3rd edn. New York State Psychiatric Hospital, New York

Stevens JD (1972) The distribution of phospholipid fractions in the red cell membrane of schizophrenics. Schizophr Bull 6: 60–61

Suddath RL, Christison GW, Torrey EF, Casanova MF, Weinberger DR (1990) Anatomical abnormalities in the brains of monozygotic twins discordant for schizophrenia. N Engl J Med 322: 789–794

Weinberger DR (1987) Implications of normal brain development for the pathogenesis of schizophrenia. Arch Gen Psychiatry 44: 660–669

Weinberger DR, Berman KF, Zec DF (1986) Physiological dysfunction of the dorsolateral prefrontal cortex in schizophrenia. Arch Gen Psychiatry 43: 114–124

Wong DF, Wagner HN, Tune LE, Dannals RF, Pearlson GD, Links JM, Tamminga CA, Broussolle EP, Ravert HT, Wilson AA, Toung JKT, Malat J, Williams JA, Lorcan A, O'Tuama O, Snyder SH, Kuhar MJ, Gjedde A (1986) Positron emission tomography reveals elevated D2 dopamine receptors in drug-naive schizophrenics. Science 234: 1558–1563

Zubenko GS, Cohen BM, Reynolds CF, Boller F, Malinakova I, Keefe MA (1987) Platelet membrane fluidity in Alzheimer's disease and major depression. Am J Psychiatry 144: 860–868

Authors' address: Dr. J. W. Pettegrew, Western Psychiatric Institute and Clinic, c/o The Graduate School of Public Health, Room A-712 Crabtree Hall, 130 DeSoto Street, Pittsburgh, PA 15261, U.S.A.

Basal ganglia

J Neural Transm (1992) [Suppl] 36: 43–59

The neostriatal mosaic: multiple levels of compartmental organization

C. R. Gerfen

Laboratory of Cell Biology, National Institute of Mental Health, Bethesda, MD, U.S.A.

Summary. Although schizophrenia may result from dysfunction of the cerebral cortex the possible indirect involvement of the basal ganglia may be important as this neural system provides a major neural system through which the cortex affects behavior. Processing of cortical input occurs within the striatum, which is the main component of the basal ganglia, where excitatory cortical imput is transformed to oppositely modulate the output nuclei of the basal ganglia. The details of this transformation, as well as the role of dopamine in this process, are beginning to unfold. Striatal projections to the globus pallidus, through connections with the subthalamic nucleus, modulate excitatory input to the output neurons of the basal ganglia, GABAergic neurons in the internal segment of the globus pallidus and in the substantia nigra, whereas striatal projections directly to these neurons, provide inhibitory inputs. Thus, cortically driven activity in these two striatal output pathways oppositely modulate the output neurons of the basal ganglia. Dopamine appears to play a crucial role in this transformation. D_1 and D_2 dopamine receptors are specifically expressed by striatonigral and striatopallidal neurons, respectively. The direct action of dopamine through these receptors appears to oppositely modulate the responsiveness of striatal output pathways to cortical input. Insights into the role of dopaminergic function within the basal ganglia may have direct relevance to the development of treatments for schizophrenia.

Introduction

At this time considerable evidence suggests that schizophrenia results from cortical dysfunction, whereas the basal ganglia may be only indirectly involved. Nonetheless, the intimate relationship between the cortex and the basal ganglia, and the possibility that pharmacologic therapeutic treatments may act, at least in part, within the basal ganglia, suggest that even if only indirect, the involvement of the basal ganglia in schizophrenia may be important. The basal ganglia provide a major neural system through which the cerebral cortex affects behavior. In this paper the neuroanatomical and neurochemical systems involved in processing cortical input by the basal ganglia will be described. The striatum is suggested to consist of multiple levels of functional compartments that are related in some respects to the laminar and regional organization of the cortex, but in other respects provide a unique form of neural processing. One level is related to the segregation of its two major output systems, one to the globus pallidus and the other to the substantia nigra. Another level relates the patch and matrix compartments of the striatum to the segregation of the cortical outputs of sublaminae of layer 5. How these functional compartments are integrated in different striatal regions provides further insights as to the functional organization of the basal ganglia processing of cortical information.

The general organization of the basal ganglia (see Wilson, 1990, for review) is diagrammed in Fig. 1. The predominant input to the basal ganglia is the excitatory input from the cerebral cortex (and from the thalamus) to the striatum. Striatal neurons are generally quiescent and activity is generated by this excitatory input. The vast majority of neurons in the striatum are medium spiny output neurons (Wilson and Groves, 1980), which constitute some 90–95% of the neuronal population. These neurons, all of which utilize GABA as a transmitter (Chevalier et al., 1985; Deniau and Chevalier, 1985; Kita and Kitai, 1988), give rise to two major output pathways (Kawaguchi et al., 1990), for the most part from separate, approximately equal numbers of neurons. One pathway, the so-called direct striatonigral output pathway, provides inputs to the entopeduncular nucleus (internal segment of the globus pallidus in primates) and the substantia nigra (Kawaguchi et al., 1990). For the purposes of this review these two nuclei are considered part of one extended nucleus and the term striatonigral refers to the pathway providing inputs to both nuclei. The second striatal output pathway provides an indirect pathway between the striatum and the substantia nigra. Striatal neurons contributing to this pathway provide inputs to the globus pallidus (the external segment of the globus pallidus in primates) (Kawaguchi et al., 1990). Neurons in the globus pallidus, which are primarily GABAergic, provide inhibitory inputs to the substantia nigra and to the subthalamic nucleus (Kita et al., 1983; Smith and Bolam, 1990; Smith et al., 1990). The subthalamic nucleus provides an excitatory input to the substantia nigra (Kita and Kitai, 1987). Thus, these two striatal output pathways provide for the opposite modulation of the substantia nigra (Albin et al., 1989). The indirect pathway, through connections with the globus pallidus and subthalamic nucleus modulate the excitatory inputs to the substantia nigra. Inhibitory inputs to the substantia nigra are regulated by the striatonigral pathway and also by the striatopallidal pathway, through inputs from the globus pallidus directly to the substantia nigra. These pathways will be discussed in more detail below.

GABAergic neurons in the substantia nigra pars reticulata and entopeduncular nuclear complex provide the major output of the basal ganglia (Chevalier et al., 1985; Deniau and Chevalier, 1985; Gerfen et al., 1982;

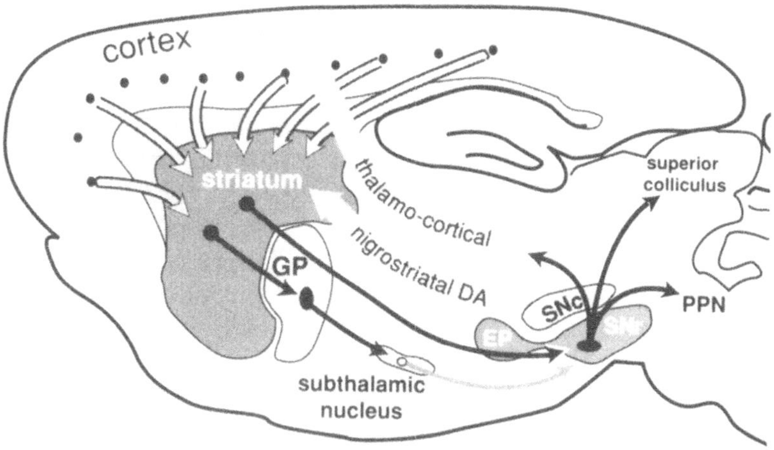

Fig. 1. Representation of the major connections of the basal ganglia. The main component of the basal ganglia, the striatum, receives inputs from the cortex. Two major striatal output pathways target the globus pallidus (*GP*) and entopeduncular (*EP*) — substantia nigra pars reticulata (*SNr*) complex. Dopamine (DA) neurons in the substantia nigra pars compacta (*SNc*) receive inputs from the striatum (not diagrammed) and provide feedback via the nigrostriatal DA pathway. EP and SNr neurons provide inhibitory inputs to the pedunculopontine nucleus (*PPN*), superior colliculus, and thalamus. Nigrothalamic inputs target intralaminar thalamic nuclei that provide feedback to the striatum (not shown) and ventral tier thalamic nuclei that provide inputs to the frontal cortex

Wilson, 1990). These neurons innervate ventral tier thalamic nuclei, which provide inputs to the frontal cortex, and to the intralaminar nuclei, which provide feedback to the striatum. In addition substantia nigra GABAergic neurons provide inputs to the superior colliculus and pedunculopontine nucleus. Dopaminergic neurons in the midbrain are located in the ventral tegmental area, the substantia nigra pars compacta, in islands embedded within the substantia nigra pars reticulata, and in the retrorubral area (Gerfen et al., 1987; Jimenez-Castellanos and Graybiel, 1987). These neurons provide feedback to the striatum and to the prefrontal cortex. The role of these neurons in the modulation of striatal output neurons will be discussed in detail below.

Striatopallidal and striatonigral systems

One level of functional compartmental organization in the striatum is related to the separate populations of striatal medium spiny neurons that provide inputs to the globus pallidus (the external segment of the globus pallidus in primates) and to the entopeduncular/substantia nigra complex (Kawaguchi et al., 1990). These two major types of neurons are referred to as striatopallidal neurons and striatonigral neurons. All medium spiny neurons have a local axon collateral that is distributed within the striatum and projection axons that extend out of the striatum (Wilson and Groves, 1980). Striatopallidal neurons project an axon with a dense arborization within the globus pallidus (Kawaguchi et al., 1990). Striatonigral neurons also extend an axon collateral within the globus pallidus, but with a reduced arborization, and additionally extend axon collaterals that arborize densely in the entopeduncular nucleus and/or substantia nigra (Kawaguchi et al., 1990). All striatopallidal and striatonigral neurons appear to utilize GABA as a neurotransmitter (Kita and Kitai, 1988). The localization of the GABA receptors in both pallidal and nigral target neurons of these projections (Lolait et al., 1989), and the GABA receptor mediated inhibition of these neurons (Chevalier et al., 1985; Deniau and Chevalier, 1985) provide further evidence of the role of GABA as the primary neurotransmitter in striatopallidal and striatonigral projections.

The numbers of striatopallidal and striatonigral neurons are approximately equal and are intermingled throughout the striatum (Gerfen and Young, 1988). While striatopallidal and striatonigral neurons share some common morphologic and neurochemical characteristics they may be distinguished not only on the basis of their projections, but also by their expression of different neuropeptides and dopamine receptor subtypes (see Fig. 2). The majority of striatopallidal neurons express the peptide enkephalin and the D2 dopamine receptor, whereas the majority of striatonigral neurons express both substance P and dynorphin and the D1 dopamine receptor (Gerfen et al., 1990; Le Moine et al., 1990). While these neurochemical determinants apply to the majority of striatopallidal and striatonigral neurons there are important exceptions. A number of striatal projection neurons co-express both the D1 and D2 dopamine receptor and a similar number co-express enkephalin and substance P (and possibly dynorphin) (Penny et al., 1986). Estimates of between 20–30% of striatal neurons have been given for such co-expression patterns. However, additional work is needed to determine whether neurons that co-express the dopamine receptor subtypes also co-express the inappropriate peptide combinations and the axonal projections of these neurons. Additionally, there are regional differences in the patterns of expression of peptides by different output neurons (Gerfen and Young, 1988), which will be discussed below.

Keeping the exceptions noted in mind, the majority of striatal output neurons constitute two major types, striatopallidal neurons that express the D2 dopamine receptor and enkephalin, and striatonigral neurons that express the D1 dopamine receptor and the peptides dynorphin and substance P (Gerfen et al., 1990; Le Moine et al., 1990). Dopamine depleting lesions and pharmacologic treatments produce specific changes in the levels of peptide mRNAs expressed by striatal output neurons that provide one assay of the dopaminergic regulation of these neurons. Depletion of striatal dopamine with 6-hydroxydopamine (6-OHDA) or treatments of animals with neuroleptics result in an elevation of enkephalin mRNA levels in striatopallidal neurons, and a reduction of substance P and dynorphin mRNA levels in striatonigral neurons (Bannon et al., 1986; Gerfen et al.,

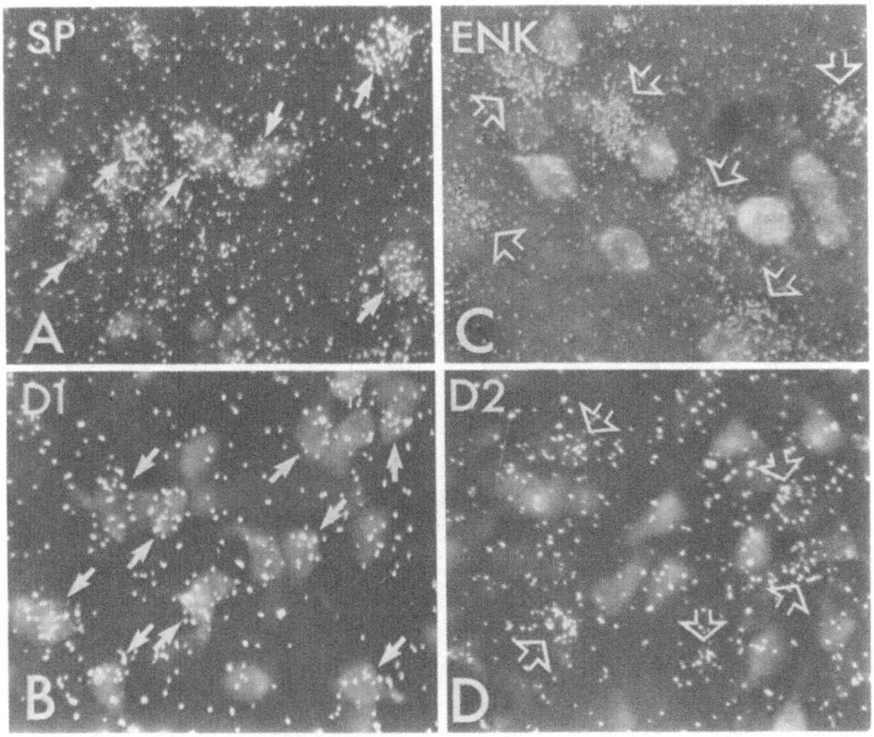

Fig. 2. Striatal neurons retrogradely labeled with the fluorescent dye fluorogold after injection into the substantia nigra combined with darkfield illumination of silver grains produced by ISHH labeling with ^{35}S-labeled oligonucleotide probes for **A** substance P (*SP*), **B** the D1 dopamine receptor (*D1*), **C** enkephalin (*ENK*) and **D** the D2 dopamine receptor (*D2*). Striatonigral neurons show ISHH labeling for both substance P (A, solid arrows) and the D1 dopamine receptor (B, solid arrows). Striatal neurons that are unlabeled by fluorgold, and presumably project to the globus pallidus, show ISHH labeling for both enkephalin (C, open arrows) and the D2 dopamine receptor (D, open arrows). From Gerfen et al. (1990)

1991; Hong et al., 1978, 1985; Li et al., 1988, 1987; Morris et al., 1988; Voorn et al., 1987; Young et al., 1986). Animals with unilateral 6-OHDA that subsequently receive semi-chronic treatment (up to 21 days) with D1 or D2 selective dopamine agonists show opposite modulation of peptide mRNA levels in striatonigral and striatopallidal neurons (Gerfen et al., 1990). In these animals D1 selective agonist treatment reverses the lesion-induced reduction in substance P mRNA and results in a significant increase in dynorphin mRNA levels in striatonigral neurons and is without effect on enkephalin mRNA levels in striatopallidal neurons, whereas D2 selective agonist treatment selectively reverses the lesion-induced increase in enkephalin mRNA levels in striatopallidal neurons. These results (see Fig. 3) suggest that the direct action of dopamine on striatonigral and striatopallidal neurons is mediated by the specific expression of the D1 and D2 dopamine receptors, respectively, by these neurons. Thus in the

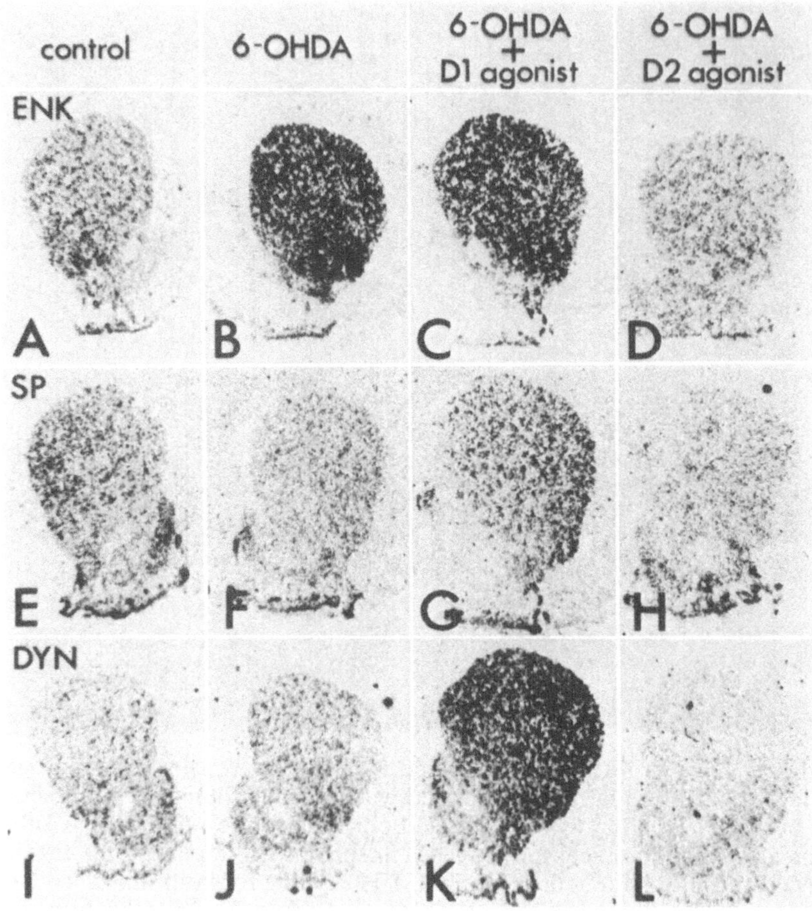

Fig. 3. In situ hybridization in the striatum from brain sections apposed to autoradiographic film with ^{35}S-labeled oligonucleotide probes complementary to (**A** through **E**) enkephalin (*ENK*), (**F** through **I**) substance P (*SP*), and (**J** through **L**) dynorphin (*DYN*). Sections in the first two columns are from the same saline treated animal showing ISHH labeling on the unlesioned control side (A,F,J) and 6-OHDA lesioned side (B,G,K). Sections in the third column are from the 6-OHDA lesioned side of an animal that received intermittent treatment with the D1 agonist SKF-38393 (C,G,K). Sections in the fourth column are from the 6-OHDA lesioned side of an animal that received continuous treatment with the D2 agonist quinpirole (D,H,L). The increase in enkephalin ISHH labeling caused by 6-OHDA lesions (B) is not affected by D1 agonist treatment (C) but is reversed by continuous D2 agonist treatment (D). The decrease in substance P ISHH labeling caused by 6-OHDA lesions (F) is reversed by intermittent D1 agonist treatment (G) but is unaffected by D2 agonist treatment (H). Dynorphin ISHH labeling is not significantly altered by 6-OHDA lesions (J) but is elevated by D1 agonist treatment (K) but not affected by D2 agonist treatment (L). From Gerfen et al. (1990)

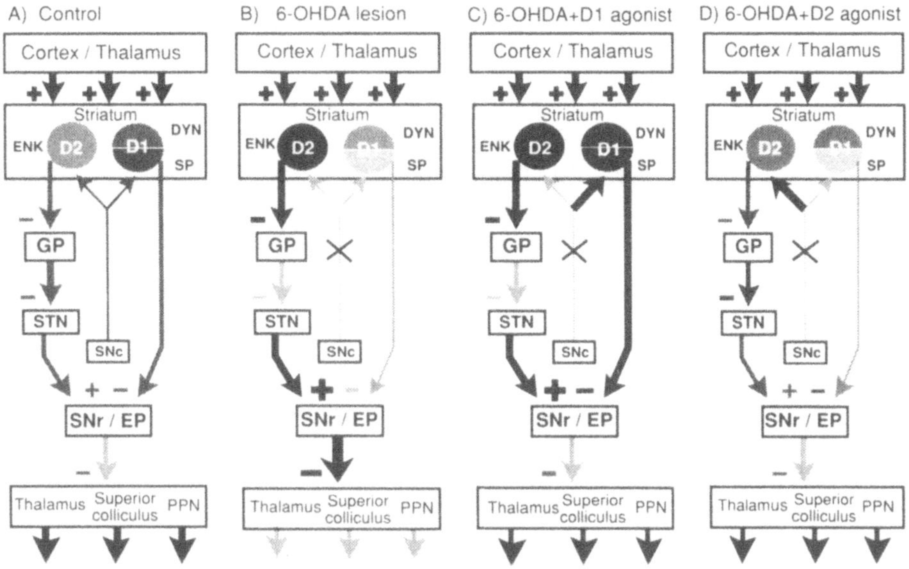

Fig. 4. Diagram of the connections of striatal output neurons. The grey level of the cells denote relative peptide mRNA levels and the thickness of the lines indicate relative activity. **A** *Control*: The cortex and thalamus provide an excitatory input to the striatum. Striatal neurons that contain enkephalin (*ENK*) and the D2 dopamine receptor (*D2*) provide an inhibitory input to the globus pallidus (*GP*). Pallidal neurons provide an inhibitory input to the subthalamic nucleus (*STN*), which provides an excitatory input to the substantia nigra pars reticulata (*SNr*). Striatal neurons that express the D1 dopamine receptor (*D1*), dynorphin (*DYN*) and substance P (*SP*) provide an inhibitory input to the substantia nigra/entopeduncular nucleus (*SNr/EP*). SNr/EP GABAergic neurons inhibit neurons in the thalamus, superior colliculus and pedunculopontine nucleus (*PPN*). Normal behavioral activity (arrows at the bottom of the diagram) is dependent on coordinated striatonigral and striatopallidal outputs that regulate SN output. **B** *6-OHDA*: Dopamine lesions result in increased enkephalin expression and activity in striatopallidal neurons. This results in increased firing of SN GABAergic neurons and in diminished behavioral activity (arrows at bottom of diagram). **C** *6-OHDA + D1 agonist*: D1 agonist treatment after 6-OHDA lesions does not alter the lesion induced increase in enkephalin in the striatopallidal pathway but reverses the decrease in substance P and significantly increases dynorphin in striatongral neurons. **D** *6-OHDA + D2 agonist*: Continuous D2 agonist treatment after 6-OHDA lesions has no effect on the striatonigral pathway but reverses the lesioned induced increase in enkephalin in the striatopallidal pathway. This reverses the increased excitatory input from the subthalamic nucleus to the substantia nigra pars reticulata. From Gerfen et al. (1990)

dopamine depleted striatum activation of D1 and D2 dopamine receptors appears to oppositely effect gene regulation of striatonigral and striatopallidal neurons, respectively.

The consequences of dopamine's actions through D1 and D2 dopamine receptors on the circuitry of the basal ganglia, determined by data on gene regulation (Gerfen et al., 1990), 2-deoxyglucose utilization (Trugman and Wooten, 1987) and electrophysiological studies (Carlson et al., 1990; Pan and Walters, 1988; Wieck and Walters, 1987) are diagrammed in Fig. 4.

Following dopamine depletion in the striatum increased activity of the striatopallidal system results in an increase in the tonic firing of nigral GABAergic neurons. In the dopamine depleted striatum D1 selective agonists specifically elevate the activity of striatonigral neurons, whereas D2 selective agonists reverse the lesion-induced increase in striatopallidal neurons. Thus it appears that the direct action of dopamine through D1 and D2 receptors is to oppositely modulate striatonigral and striatopallidal neurons, respectively.

The behavioral consequences of opposed modulation of GABAergic nigral neurons are generally modeled on the concept that the tonic inhibitory output of these neurons, generated in part by input from the subthalamic nucleus, is interrupted by inhibitory inputs from the striatum during movements (Chevalier et al., 1985; Deniau and Chevalier, 1985). This disinhibitory process has been best characterized for eye movements by Hikosaka and Wurtz (1983a,c,d) who in a series of elegant studies showed that visual, memory and reward contingent eye movements were correlated with pauses in the tonic activity of nigral GABAergic neurons and increased activity in superior colliculus neurons. Akinesia which accompanies Parkinson's disease has been related to increased striatopallidal activity and the resultant increase in excitatory subthalamic inputs to nigral GABAergic neurons, which then predominates over the disinhibitory mechanisms required for the generation of movements. This model is supported by the report that lesions of the subthalamic nucleus reverse akinesia in monkeys made Parkinsonian with MPTP-induced lesions (Bergman et al., 1990).

The model presented of dopaminergic modulation of the striatopallidal and striatonigral systems is oversimplified. The specific action through D1 and D2 receptors respectively on striatonigral and striatopallidal outputs reflects the direct action in the unique situation when one receptor is activated in the absence of activation of the other. However, in the normally innervated striatum co-activation of D1 and D2 dopamine receptors in concert with the actions of other neurotransmitters provide a myriad of interactions between striatal neurons. Synergistic effects of D1 and D2 dopamine receptors have been well documented (Carlson et al., 1985; Clark and White, 1987; Walters et al., 1987). The cellular basis of this synergism has not been established but at least three likely mechanisms might be postulated. First, as mentioned, approximately 20% of striatal projection neurons express both D1 and D2 dopamine receptors in which synergism may occur in individual neurons. Second, there may be interactions between neurons expressing different receptor subtypes through local axon collaterals (Park et al., 1980; Wilson and Groves, 1980). Third, striatal interneurons may mediate interactions between D1 and D2 expressing neurons. An example of this type of interaction may involve acetylcholine modulation of D2 containing striatopallidal neurons, acetylcholine release being induced via substance P (Arenas et al., 1991) released by D1 containing striatonigral neurons acting on substance P receptors expressed by cholinergic neurons (Gerfen, 1991). These hypothesized mechanisms

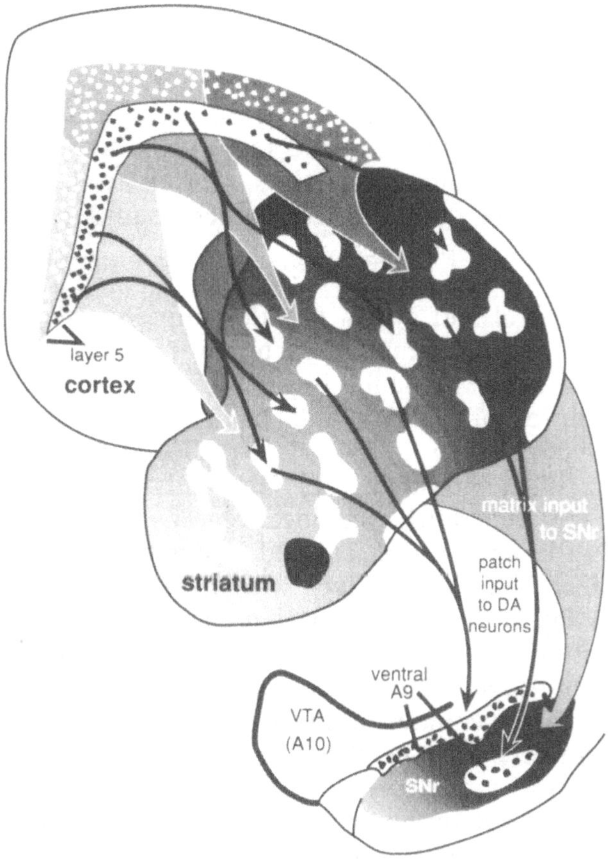

Fig. 5. Summary of patch-matrix compartmental organization of corticostriatal and striatonigral pathways. Corticostriatal neurons in the deep parts of layer 5 provide inputs to the striatal patch compartment, whereas superficial layer 5 neurons provide inputs to the striatal matrix. Patch neurons provide inputs to the location of dopaminergic neurons in the ventral tier of the substantia nigra pars compacta and islands of dopamine neurons in the pars reticulata. Striatal matrix neurons provide inputs to the location of GABAergic neurons in the substantia nigra pars reticulata (*SNr*)

mediating interactions amongst striatal neurons are by no means exhaustive. The direct action of dopamine on D1 and D2 containing striatonigral and striatopallidal neurons should be considered an important, but not the only, mechanism by which dopamine modulates the balance in the activity of these two striatal output pathways.

Patch-matrix striatal compartments

A second level of functional compartmental organization in the striatum is the segregation of medium spiny neurons into patch and matrix compartments (see Fig. 5), which have distinct projection targets, respectively providing inputs to the location of dopaminergic and GABAergic neurons in the substantia nigra (Gerfen, 1984, 1985; Gerfen et al., 1985; Jimenez-Castellanos and Graybiel, 1989). The dendrites of patch and matrix projection neurons are restricted, for the most part, to the compartment of their parent cell body (Bolam et al., 1988; Gerfen et al., 1985; Herkenham et al., 1984; Kawaguchi et al., 1989), which maintains a segregation of the inputs that are compartmentally organized. Striatal afferents that are organized with respect to the patch-matrix compartments arise from separate sets of ventral and dorsal tier dopaminergic neurons in the midbrain (Gerfen et al., 1987; Jimenez-Castellanos and Graybiel, 1987), from different subnuclei of the thalamic intralaminar complex (Berendse et al., 1988; Herkenham et al., 1984) and from the cortex (Donoghue and Herkenham, 1986; Gerfen, 1984, 1989; Goldman and Nauta, 1977; Goldman-Rakic, 1982; Regsdale and Graybiel, 1981). Perhaps most revealing as to the functional significance of the striatal patch-matrix compartments is the finding that cortical inputs to the patch compartment originate from deep layer V and layer VI, whereas those to the matrix arise from superficial layer V and supragranular layers (Gerfen, 1989). The laminar organization of the cortex reflects the segregation of neurons with common projection targets into discrete layers (or sublayers) (Jones, 1984). Thus the patch-matrix compartments of the striatum appear to provide parallel input-output pathways through which the segregated output of sublaminae of layer V of the cortex is maintained to differentially affect the dopaminergic and GABAergic neurons in the substantia nigra (Gerfen, 1984, 1989). In other words, it appears that the striatal patch-matrix compartments are related to the laminar organization of the cortex.

Here patch-matrix compartments are characterized by input-output organization although historically they were first described with neurochemical markers. In the adult striatum, the first indication of patch-matrix organization was the report of enriched distribution of μ-opiate receptors in patches (Pert et al., 1976), later Graybiel and Ragsdale reported that acetylcholinesterase staining was sparse in zones they termed striosomes (Graybiel and Ragsdale, 1978), which exactly coincide with the μ-opiate receptor patches (Herkenham and Pert, 1981). These markers, together with somatostatin fibers (Chesselet and Graybiel, 1985; Gerfen, 1985) and calbindin immunoreactive neurons (Gerfen et al., 1985), which are both specific to the matrix, display consistent patch-matrix distributions in both the dorsal and ventral striatum, with the notable exception of the shell region of the nucleus accumbens. The input-output relationships of the patch-matrix compartments described above (Gerfen, 1984, 1985, 1989; Gerfen et al., 1987) have been charted with respect to these neurochemical markers. Particularly useful is calbindin immunoreactivity which is specifically localized in matrix neurons and in their axon terminals, which confirms the matrix specific input to the location of GABAergic neurons in the substantia nigra and not to the location of dopamine neurons (Gerfen et al., 1985). These studies have established the morphologic basis of patch-matrix compartmental organization. Other neurochemical markers, such as the immunohistochemical localization of enkephalin, substance P and dynorphin, show patterns of distribution that in some regions coincide with patch-matrix organization, but in other regions do not (Graybiel et al., 1981). As discussed later, these peptides are better considered in the context of another level of functional compartmental organization and their regional variation results from interactions between multiple levels of functional compartments.

The striatal patch-matrix organization has been related to limbic and non-limbic functions on the basis of inputs to the patches from limbic-related areas, such as the amygdala (Ragsdale and Graybiel, 1988) and prelimbic cortex (Donoghue and Herkenham, 1986; Gerfen, 1984), and to the matrix from sensorimotor cortical areas (Donoghue and Herkenham, 1986; Gerfen, 1984). However, such relationships reflect a transition in the relative contribution of patch and matrix projecting neurons from periallocortical to neocortical areas. For example neocortical areas provide inputs to the patch compartment that are greatly diminished compared to those from periallocortical areas (Ferino et al., 1987; Gerfen, 1989). Rather than attempt to define the type of information conveyed to the striatal patch-matrix compartments in limbic and sensorimotor terms, it is suggested that regional transitions in both cortical and basal ganglia organization be examined. Alheid and Heimer (1988) have proposed that the subcortical connections of allo- and neocortical areas share common general organizational schemes whose specific elements reflect a transition in the final targets of these systems and in the feedback mechanisms they employ. The regional transitions in the basal ganglia may be considered in this context. The main outputs of the ventral striatum target both cholinergic (Grove et al., 1986) and dopaminergic neurons (Nauta et al., 1978), two neurochemical systems that provide direct feedbacks to the cortex and striatum. Conversely, the main outputs of the dorsal striatum target GABAergic neurons in the globus pallidus and substantia nigra, and through the thalamic connections of the latter, provide a more indirect feedback to the cortex and striatum (Wilson, 1990). As diagrammed in Fig. 6, it is suggested that these two types of circuits are contained in the connections of the patch and matrix compartments, respectively (Gerfen, 1984, 1985). In many regions of the striatum both of these circuits exist. Rather than specify that the patch compartment conveys limbic information in dorsolateral patches it is suggested that the type of neuroanatomical circuitry that is typical of the "limbic-ventral" striatum, is retained in the dorsal striatum in the connections of the patch compartment. This is analogous to the emergence of neocortical organization in which the "non-specific" connections of older cortical areas are retained to some extent but supplanted by new "specific" connections (Herkenham, 1986). Thus striatal patch-matrix compartments may be viewed as two phylogenetically distinct neuroanatomical circuits through which cortical information is processed. Regionally, the mix of these two circuit systems vary such that in the ventromedial striatum

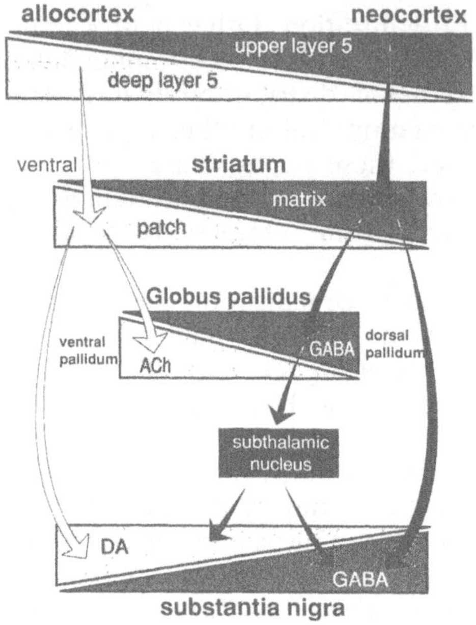

Fig. 6. Diagrammatic representation of regional transitions in the components of patch-matrix compartmental connections. See text for details. *ACh* acetylcholine, *DA* dopamine

allo- and periallocortical circuitry dominates, whereas in the dorsolateral striatum neocortical circuitry dominates. In much of the striatum the two circuits coexist and interactions between them may provide mechanisms for regulating the balance in the striatopallidal and striatonigral systems.

Integration between functional compartments

The two functional compartments described above are composed of the same populations of medium spiny neurons. They are considered distinct from the perspective of their respective roles in processing cortical information. As discussed, striatopallidal and striatonigral pathways transform excitatory cortical inputs into parallel striatal output pathways that provide antagonistic regulation of the output pathways of the basal ganglia. Several of the cellular and molecular mechanisms that modulate the relative output of these pathways have been introduced. These include the specific expression of D1 and D2 dopamine receptor subtypes by striatonigral and striatopallidal neurons, a subpopulation of striatal output neurons that express both receptor subtypes, local interactions through axon collaterals between striatal output neurons, and involvement of striatal intrinsic neurons, such as the cholinergic neurons. Such mechanisms provide for complexly diversified patterns of modulation of striatal output pathways to regulate the GABAergic output neurons of the internal segment of the globus pallidus and substantia nigra.

It is suggested that the patch-matrix compartmental organization provides yet another mechanism for regulating striatonigral and striatopallidal output pathways. The specific projection of the striatal patch compartment to the dopamaninergic nigrostriatal neurons (Gerfen, 1984, 1985; Gerfen et al., 1985; Jimenez-Castellanos and Graybiel, 1989) provides for the integration between the two functional compartments described. This pathway provides for the potential regulation by the patch compartment of the dopaminergic nigrostriatal feedback pathway, which as discussed above plays an important role in modulating the relative balance in the striatonigral and striatopallidal output pathways. Details of how this integration occur remain to be worked out, however, it would appear that there are distinct regional differences. These mechanisms appear to differ regionally within the striatum. For example, the levels of various neuroactive peptides in striatal output neurons is regulated by dopaminergic mechanisms. Although substance P and dynorphin mRNAs are expressed by approximately equal numbers of patch and matrix neurons, the relative expression by neurons in these two compartments vary regionally (Gerfen and Young, 1988; Graybiel et al., 1981). Dynorphin mRNA is expressed at higher levels by patch versus matrix neurons in the dorsal striatum, whereas in the ventral striatum the relative levels are approximately equal in the two compartments. Conversely, substance P mRNA shows an inverse pattern of expression compared to that of dynorphin, being expressed at higher levels in patch neurons in the ventral versus dorsal striatum. These differences in distribution may be related to the regional transition in the organization of the corticostriatal input and striatal organization discussed above. Thus, these peptides, rather than reflecting a strict patch-matrix organization, display a distribution that reveals the interaction between the striatal functional compartments. If, as suspected, schizophrenia selectively involves specific cortical areas, then the regional differences in the mechanisms regulating striatal output systems could be of some importance.

In concluding one last observation concerning the dopaminergic mechanisms regulating striatal output pathways may be of some importance. The direct action of D1 and D2 dopamine selective agonists in regulating striatonigral and striatopallidal pathways depends on the time course of drug administration. To be effective, D2 selective agonists required continuous administration, whereas D1 selective agonists produced marked changes with intermittent treatment schedules (Gerfen et al., 1990). Grace (1991) has suggested that prefrontal hypofunction in schizophrenia results in a disequilibrium in the normal mechanisms regulating phasic and tonic release of dopamine within the striatum. Given the marked differences in the regulation of striatal output neurons to continuously and intermittently administered pharmacologic treatments it would seem that some attention should be given to the schedule of treatment in the pharmacologic intervention in schizophrenia.

References

Albin RL, Young AB, Penney JB (1989) The functional anatomy of basal ganglia disorders. Trends Neurosci 12: 366–375

Alheid GF, Heimer L (1988) New perspectives in basal forebrain organization of special relevance for neuropsychiatric disorders: the striatopallidal, amygdaloid, and corticopetal components of the substantia innominata. Neuroscience 27: 1–39

Arenas E, Alberch J, Perez-Navarro E, Solsona C, Marsal J (1991) Neurokinin receptors differentially mediate endogenous acetycholine release evoked by tachykinins in the neostriatum. J Neurosci 11:2332–2338

Bannon MJ, Lee J-M, Girand P, Young A, Affolter J-U, Bonner TI (1986) Dopamine antagonist haloperidol decreases substance P, substance K and preprotachykinin mRNAs in rat striatonigral neurons. J Biol Chem 261: 6640–6642

Berendse HW, Voorn P, Kortschot AT, Groenewegen HJ (1988) Nuclear origin of thalamic afferents of the ventral striatum determines their relation to patch/matrix configurations in enkephalin immunoreactivity in the rat. J Chem Neuroanat 1: 3–10

Bergman H, Whitman T, DeLong MR (1990) Reversal of experimental parkinsonism by lesions of the subthalamic nucleus. Science 249: 1436–1438

Bolam JP, Izzo IN, Graybiel AM (1988) Cellular substrate of the histochemically defined striosome/matrix system of the caudate nucleus: a combined Golgi and immunocytochemical study in cat and ferret. Neuroscience 24: 853–875

Carlson JH, Bergrstrom DA, Walters JR (1985) Stimulation of both D1 and D2 dopamine receptors appears as necessary for full expression of postynapitc effects of dopamine agonists: a neurophysiological study. Brain Res 400: 205–218

Carlson JH, Bergstrom DA, Demo SD, Walters JR (1990) Nigrostriatal lesion alters neurophysiological responses to selective and nonselective D-1 and D-2 dopamine agonists in rat globus pallidus. Synapse 5: 83–93

Chesselet M-F, Graybiel AM (1985) Striatal neurons expressing somatostatin-like immunoreactivity: evidence for a peptidergic interneuronal system in the cat. Neuroscience 17: 547–571

Chevalier G, Vacher S, Deniau JM, Desban M (1985) Disinhibition as a basic process in the expression of striatal function. I. The striato-nigral influence on tecto-spinal/tecto-diencephalic neurons. Brain Res 334: 215–226

Clark D, White FJ (1987) Review: D1 dopamine receptor-the search for a function: a critical review of the D1/D2 dopamine receptor classification and its functional implications. Synapse 1: 347–388

Deniau JM, Chevalier G (1985) Disinhibition as a basic process in the expression of striatal functions. II. The striato-nigral influence on thalamocortical cells of the ventromedial thalamic nucleus. Brain Res 334: 227–233

Donoghue JP, Herkenham M (1986) Neostriatal projections from individual cortical fields conform to histochemically distinct striatal compartments in the rat. Brain Res 365: 397–403

Ferino F, Thierry AM, Saffroy M, Glowinski J (1987) Interhemispheric and subcortical collaterals of medial prefrontal cortical neurons in the rat. Brain Res 417: 257–266

Gerfen CR (1984) The neorstriatal mosaic: compartmentalization of corticostriatal input and striatonigral output systems. Nature 311: 461–464

Gerfen CR (1985) The neostriatal mosaic. I. Compartmental organization of projections from the striatum to the substantia nigra in the rat. J Comp Neurol 236: 454–476

Gerfen CR (1989) The neostriatal mosaic: striatal patch-matrix organization is related to cortical lamination. Science 246: 385–388

Gerfen CR (1991) Substance P (Neurokinin-1) receptor mRNA is selectively expressed in cholinergic neurons in the striatum and basal forebrain. Brain Res 556:165–170

Gerfen CR, Young WS (1988) Distribution of striatonigral and striatopallidal peptidergic neurons in both patch and matrix compartments: an in situ hybridization histochemistry and fluorescent retrograde tracing study. Brain Res 460: 161–167
Gerfen CR, Staines WA, Arbuthnott GW, Fibiger HC (1982) Crossed connections of the substantia nigra in the rat. J Comp Neurol 207: 283–303
Gerfen CR, Baimbridge KG, Miller JJ (1985) The neostriatal mosaic: compartmental distribution of calcium binding protein and parvalbumin in the basal ganglia of the rat and monkey. Proc Natl Acad Sci USA 82: 8780–8784
Gerfen CR, Baimbridge KG, Thibault J (1987) The neostriatal mosaic. III. Biochemical and developmental dissociation of dual nigrostriatal dopaminergic systems. J Neurosci 7: 3935–3944
Gerfen CR, Herkenham M, Thibault J (1987) The neostriatal mosaic. II. Patch- and matrix-directed mesostriatal dopaminergic and non-dopaminergic systems. J Neurosci 7: 3915–3934
Gerfen CR, Engber TM, Mahan LC, Susel Z, Chase TN, FJ Monsma J, Sibley DR (1990) D1 and D2 dopamine receptor-regulated gene expression of striatonigral and striatopallidal neurons. Science 250: 1429–1432
Gerfen CR, McGinty JF, Young WS (1991) Dopamine differentially regulates dynorphin, substance P and enkephalin expression in striatal neurons: in situ hybridization histochemical analysis. J Neurosci 11: 1016–1031
Goldman PS, Nauta WJH (1977) An intricately patterned prefronto-caudate projection in the rhesus monkey. J Comp Neurol 171: 369–385
Goldman-Rakic PS (1982) Cytoarchitectonic heterogeneity of the primates neostriatum: subdivision into island and matrix cellular compartments. J Comp Neurol 205: 398–413
Grace AA (1991) Phasic versus tonic dopamine release and the modulation of dopamine system responsivity: a hypothesis for the etiology of schizophrenia. Neuroscience 41: 1–24
Graybiel AM, Ragsdale CW (1978) Histochemically distinct compartments in the striatum of human, monkey and cat demonstrated by acetylcholinesterase staining. Proc Natl Acad Sci USA 75: 5723–5726
Graybiel AM, Ragsdale CWJ, Yoneika ES, Elde RP (1981) An immunohistochemical study of enkephalins and other neuropeptides in the striatum of the cat with evidence that the opiate peptides are arranged to form mosaic patterns in register with the striosomal compartments visible with acetylcholinesterase staining. Neuroscience 6: 377–397
Grove EA, Domesick VB, Nauta WJH (1986) Light microscopic evidence of striatal input to intrapallidal neurons of cholinergic cell group Ch4 in the rat: a study employing the anterograde tracer Phaseolus vulgaris leucagglutinin (PHA-L). Brain Res 367: 379–384
Herkenham M (1986) New perspectives on the organization and evolution of nonspecific thalamocortical projections. In: Jones EG, Peters A (eds) Cerebral cortex. Plenum, New York, pp 403–445
Herkenham M, Pert CB (1981) Mosaic distribution of opiate receptors, parafascicular projections and acetylcholinesterase in rat striatum. Nature (London) 291: 415–418
Herkenham M, Edley SM, Stuart J (1984) Cell clusters in the nucleus accumbens of the rat, and the mosaic relationship of opiate receptors, acetylcholinesterase and subcortical afferent terminations. Neuroscience 11: 561–593
Hikosaka O, Wurtz RH (1983a) Visual and oculmotor functions of monkey substantia nigra pars reticulata. I. Relation to visual and auditory responses to saccades. J Neurophysiol 49: 1230–1253
Hikosaka O, Wurtz RH (1983c) Visual and oculmotor functions of monkey substantia nigra pars reticulata. III. Memory contingent visual and saccade responses. J Neurophysiol 49: 1268–1284

Hikosaka O, Wurtz RH (1983d) Visual and oculmotor functions of monkey substantia nigra pars reticulata. IV. Relation of subsantia nigra to superior colliculus. J Neurophysiol 49: 1285–1301
Hong JS, Yang H-YT, Costa E (1978) Substance P content of substantia nigra after chronic treatment with antischizophrenic drugs. Neuropharmacology 17: 83–85
Hong JS, Yoshikawa K, Kanamatsu T, Sabol SL (1985) Modulation of striatal enkephalinergic neurons by antipsychotic drugs. Fed Proc 44: 2535–2539
Jimenez-Castellanos J, Graybiel AM (1987) Subdivisions of the dopamine-containing A8-A9-A10 complex identified by their differential mesostriatal innervation of striosomes and extrastriosomal matrix. Neuroscience 23: 223–242
Jimenez-Castellanos J, Graybiel AM (1989) Compartmental origins of striatal efferent projections in the cat. Neuroscience 32: 297–321
Jones EG (1984) Laminar distribution of cortical efferent cells. In: Jones EG, Peters A (eds) Cerebral cortex, vol 1. Cellular components of the cerebral cortex. Plenum, New York, pp 521–553
Kawaguchi Y, Wilson CJ, Emson P (1989) Intracellular recording of identified neostriatal patch and matrix spiny cells in a slice preparation preserving cortical inputs. J Neurophysiol 62:1052–1068
Kawaguchi Y, Wilson CJ, Emson P (1990) Projection subtypes of rat neostriatal matrix cells revealed by intracellular injection of biocytin. J Neurosci 10: 3421–3438
Kita H, Kitai ST (1987) Efferent projections of the subthalamic nucleus in the rat: light and electron microscopic analysis with the PHA-L method. J Comp Neurol 260: 435–452
Kita H, Kitai ST (1988) Glutamate decarboxylase immunoreactive neurons in rat neostriatum: their morphological and populations. Brain Res 447: 346–352
Kita H, Chang HT, Kitai ST (1983) Pallidal inputs to subthalamus: intracellular analysis. Brain Res 264: 255–265
Le Moine C, Normand E, Guitteny AF, Fouque B, Teoule R, Bloch B (1990) Dopamine receptor gene expression by enkephalin neurons in rat forebrain. Proc Natl Acad Sci USA 87: 230–234
Li SJ, Sivam SP, McGinty JF, Huang YS, Hong JS (1987) Dopaminergic regulation of tachykinin metabolism in the striatonigral pathway. J Pharmacol Exp Ther 243: 792–798
Li SJ, Sivam SP, McGinty JF, Douglass J, Calavetta L, Hong JS (1988) Regulation of the metabolism of striatal dynorphin by the dopaminergic system. J Pharmacol Exp Ther 246: 403–408
Lolait SJ, O'Carroll A-M, Kusano K, Mahan LC (1989) Pharmacological characterization and region-specific expression in brain of the B2- and B3-subunits of the rat $GABA_A$ receptor. FEBS Lett 258: 17–21
Morris BJ, Remer S, Hollt V, Herz A (1988) Regulation of striatal prodynorphin mRNA levels by the raphe-striatal pathway. Mol Brain Res 4: 15–22
Nauta WJH, Smith GP, Faull RLM, Domesick VB (1978) Efferent connections and nigral afferents of the nucleus accumbens septi in the rat. Neuroscience 3: 385–401
Pan HS, Walters JR (1988) Unilateral lesion of the nigrostriatal pathway decreases firing rate and alters the firing pattern of globus pallidus neurons in the rat. Synapse 2: 650–656
Park MR, Lighthall JW, Kitai ST (1980) Recurrent inhibition in the rat neostriatum. Brain Res 194: 359–369
Penny GR, Afsharpour S, Kitai ST (1986) The glutamate decarboxylase-, leucine enkephalin-, methionine enkephalin- and substance P-immunoreactive neurons in the neostriatum of the rat and cat: evidence for partial population overlap. Neuroscience 17: 1011–1045
Pert CB, Kuhar MJ, Snyder SH (1976) Opiate receptor: autoradiographic demonstration of localization in rat brain. Proc Natl Acad Sci USA 73: 3729–3733

Ragsdale CWJ, Graybiel AM (1981) The fronto-striatal projection in the cat and monkey and its relationship to inhomogeneities established by acetylcholinesterase histochemistry. Brain Res 208: 259–266
Ragsdale CWJ, Graybiel AM (1988) Fibers from the basolateral nucleus of the amygdala selectively innervate striosomes in the caudate mucleus of the cat. J Comp Neurol 269: 506–522
Smith Y, Bolam JP (1990) The output neurones and the dopaminergic neurones of the substantia nigra receive a GABA-containing input from the globus pallidus in the rat. J Comp Neurol 296: 47–64
Smith Y, Bolam JP, Krosigk Mv (1990) Topographical and synaptic organization of the GABA-containing pallidosubthalamic projection in the rat. Eur J Neurosci 2: 500–511
Trugman JM, Wooten GF (1987) Selective D1 and D2 dopamine agonists differentially alter basal ganglia glucose utilization in rats with unilateral 6-hydroxydopamine substantia nigra lesions. J Neurosci 7: 2927–2935
Voorn P, Roest G, Groenewegen HJ (1987) Increase of enkephalin and decrease of substance P immunoreactivity in the dorsal and ventral striatum of the rat after midbrain 6-hydroxydopamine lesions. Brain Res 412: 391–396
Walters JR, Bergstrom DA, Carlson JH, Chase TN, Braun AR (1987) D1 dopamine receptor activation required for postsynaptic expression of D2 agonist affects. Science 236: 719–722
Wieck BG, Walters JR (1987) Effects of D1 and D2 dopamine receptor stimulation on the activity of substantia nigra pars reticulata neurons in 6-hydroxydopamine lesioned rats: D1/D2 coactivation induces potentiated responses. Brain Res 405: 234–246
Wilson CJ (1990) The basal ganglia. In: Shepard GM (ed) The synaptic organization of the brain. Oxford Press, New York, pp 279–316
Wilson CJ, Croves PM (1980) Fine structure and synaptic connections of the common spiny neuron of the rat neostriatum: a study employing intracellular injection of horseradish peroxidase. J Comp Neurol 194: 599–615
Young WS III, Bonner TI, Brann MR (1986) Mesencephalic dopaminergic neurons regulate the expression of neuropeptide mRNAs in the rat forebrain. Proc Natl Acad Sci USA 83: 9827–9831

Author's address: Dr. C. R. Gerfen, Laboratory of Cell Biology, National Institute of Mental Health, Bldg.36, Rm.2D-10, Bethesda, MD 20892, U.S.A.

J Neural Transm (1992) [Suppl] 36: 61–89

The regulation of subcortical dopamine systems by the prefrontal cortex: interactions of central dopamine systems and the pathogenesis of schizophrenia

A. Y. Deutch

Department of Psychiatry, Yale University School of Medicine, New Haven, CT, U.S.A.

Summary. See page 80

A recent hypothesis of the pathogenesis of schizophrenia posits a developmentally-specific dysfunction of the dopaminergic innervation of the prefrontal cortex (PFC; Weinberger, 1987; Berman and Weinberger, 1990). It has been difficult to reconcile this hypothesis with the observation that all clinically effective antipsychotic drugs used for the treatment of schizophrenia block dopamine D_2 receptors (see Deutch et al., 1991a). A resolution between the suggestion of functional dopamine (DA) "depletion" in the PFC and enhanced subcortical DA function was offered by studies of Carter, Pycock, and associates (Carter and Pycock, 1980; Pycock et al., 1980a,b). These investigators reported that depletion of DA in the rat PFC enhanced DA utilization in subcortical sites such as the nucleus accumbens septi (NAS) and striatum. Thus, a functional deficit in DA neurotransmission in the PFC would increase subcortical DA turnover, and the D_2 receptor blockade induced by antipsychotic drugs would counteract the increase in dopaminergic tone in subcortical sites. This hypothesis has been particularly influential because it incorporates both an explanation for negative symptoms, which are thought to reflect cortical dysfunction (a derangement in DA transmission in the PFC), and the efficacy of antipsychotic drugs in the treatment of positive symptoms (arising from increases in subcortical DA tone).

As attractive as this hypothesis has been, the physiological underpinnings that subserve such system interactions have remained elusive. Pycock, Carter, and colleagues (Carter and Pycock, 1980; Pycock et al., 1980a,b) reported that 6-hydroxydopamine (6-OHDA) lesions of the PFC increase DA levels and DA turnover in the striatum; certain aspects of their findings have been confirmed (Martin-Iversen et al., 1986; Leccese and Lyness, 1987; Haroutounian et al., 1988). However, other groups have been unable to confirm either the biochemical or behavioral findings of Pycock and associates (Joyce et al., 1983; Oades et al., 1986; Deutch et al., 1990).

Moreover, Pycock and colleagues did not observe consistent effects of PFC DA deafferentation on various indices of subcortical DA function (Carter and Pycock, 1980; Pycock et al., 1980a,b). In light of the importance that such DA system interactions may have in the pathogenesis of schizophrenia, we have reinvestigated the effects of cortical DA lesions on subcortical DA function.

The effects of prefrontal cortical DA depletion on subcortical DA function

The means whereby changes in the function of the PFC DA innervation may impact on subcortical DA systems have been the focus of intense interest over the past several years, in large part because of the potential insights such mechanisms hold for understanding the pathophysiology of certain symptoms in schizophrenia. Considerable advances have been made, both in terms of defining the ability of the PFC to transsynaptically regulate subcortical DA systems, and in terms of defining the mechanisms that subserve these system interactions.

Basic mechanisms

The mechanisms by which alterations in cortical DA function may impact on subcortical DA systems have become more clear over the past decade. Anatomical, pharmacological, and physiological data have led to the following hypothesized scheme by which functional decreases in the PFC DA innervation can enhance subcortical DA function.

Dopaminergic axons in the prefrontal cortex form symmetric synapses with deep layer pyramidal neurons (Seguela et al., 1988; Goldman-Rakic et al., 1989), suggesting that DA may inhibit the activity of PFC neurons that project to subcortical sites. Electrophysiological findings are congruent with the anatomical data: iontophoretic application of DA to cortical neurons antidromically activated from the striatum or thalamus inhibits the activity of these neurons (Sesack and Bunney, 1989), as does stimulation of the ventral tegmental area (VTA; Thierry et al., 1986, 1990; Peterson et al., 1987), the source of DA afferents to the PFC. The DA-induced inhibition of corticostriatal neurons appears to be mediated through a D_2 receptor isoform with atypical pharmacologic properties (Thierry et al., 1986; Sesack and Bunney, 1989). In addition to the direct DA-mediated inhibition of corticofugal neurons, DA may indirectly inhibit PFC pyramidal neurons by facilitating GABA release from interneurons (Penit-Soria et al., 1987; Retaux et al., 1991).

Cortical projection neurons predominantly use an excitatory amino acid (EAA) as a transmitter (Spencer, 1976; Kim et al., 1977; McGeer et al., 1977). Removal of these cortical neurons from tonic DA inhibition would serve to increase release of the EAA from corticostriatal nerve terminals.

We will hereafter refer to the specific EAA as glutamate, but other amino acids (including aspartate, N-acetyl-aspartylglutamate, and sulfur-containing amino acids) may also be present and potentially represent endogenous ligands for specific receptor sites. The cortical projections to the striatum form part of a triadic arrangement with striatal neurons and dopaminergic afferents, such that corticostriatal terminals abut onto the head of dendrites of striatal neurons, whereas the DA afferents form synaptic contacts with dendritic shafts (Bolam, 1984; Bouyer et al., 1984). Thus, striatal dopaminergic afferents are situated to regulate corticostriatal transmission. However, synaptic contacts between axons of cortical neurons and striatal DA axons are rare.

Glutamate has been reported to evoke DA release in the striatum via an impulse-independent (TTX-insensitive) mechanism (Roberts and Anderson, 1979; Chesselet et al., 1982; Cheramy et al., 1986; Jhamandas and Marien, 1987; Barbieto et al., 1990); recent data suggest that this may occur via reversal of the DA transporter (Lonnart and Zigmond, 1991). The specific type of EAA receptor involved is not clear. Glutamate elicits impulse-independent DA release by acting through a uniform kainate/quisqualate (AMPA family) type receptor that undergoes rapid desensitization (Barbieto et al., 1990). An NMDA-type EAA receptor may also be involved by subserving glutamate-elicited DA release which occurs through an intervening striatal neuron and is therefore TTX-sensitive. More recently, the presence of a presynaptic NMDA receptor on DA terminals that mediate glutamate-evoked DA release has been proposed (Krebs et al., 1991).

In summary, the above data suggest the presence of the requisite anatomy and pharmacology through which a decrease in cortical DA concentrations may effect increased subcortical DA function: removal of corticostriatal projection neurons from tonic DA inhibition would serve to increase glutamate release in the striatal complex and thereby evoke DA release in the subcortical site.

Effects of prefrontal cortical 6-hydroxydopamine lesions on basal striatal DA function

In attempts to replicate the findings of Carter, Pycock, and associates, adult male rats were subjected to 6-OHDA induced lesions of the anteromedial PFC. All animals were pretreated with desmethylimipramine (DMI) to minimize concomitant noradrenergic depletion. These lesions resulted in moderately complete disruption of the dopaminergic innervation of the PFC, as reflected by biochemical and anatomical measurements. For example, such lesions resulted in an 80% reduction in PFC DA concentrations; cortical NE concentrations were also significantly decreased, albeit to a lesser degree (40%; Deutch et al., 1990). There was no decrease in nucleus accumbens DA levels. Immunohistochemical analyses of the lesion revealed the loss of DA axons in the dorsomedial precentral, anterior cingulate (area 24b), prelimbic (area 32), and dorsal infralimbic (area 25) fields of the medial PFC, without any detectable involvement of the rostral nucleus accumbens or dorsal striatal DA innervations (Rosin et al., 1992).

We have been uniformly unsuccessful in attempts to replicate the findings of Pycock, Carter, and associates. PFC DA depletion induced by local 6-OHDA infusion did not increase DA metabolite concentrations in the dorsal or ventral striatum, or in the midbrain DA cell body regions; similarly, DA turnover was not enhanced in the subcortical sites (Deutch et al., 1990; Rosin et al., 1992). Measures of DA synthesis did not change following cortical DA depletion: DOPA accumulation following decarboxylase inhibition was not increased in the subcortical sites, nor was tyrosine hydroxylase (TH) enzyme activity enhanced (Rosin et al., 1992). Thus, a variety of measures of DA metabolism indicated that cortical DA depletion did not enhance increase DA function.

We were, however, intrigued by the fact that in almost every experiment there was a tendency for the dependent measure of DA function in the dorsal striatum to increase in the lesioned animals; in no case did these increases achieve a level of statistical significance. This may reflect a subthreshold increase in subcortical DA function following PFC DA depletion. Alternatively, the consistent (albeit not significant) increase in DA utilization in the dorsal striatum may reflect the fact that the prefrontal cortex innervates a relatively discrete striatal sector (Sesack et al., 1989; McGeorge and Faull, 1989), but our dissections were of the entire striatum. In order to assess if there was a regionally heterogeneous increase in dorsal striatal DA function following PFC DA lesions, we separately analyzed the medial periventricular striatum (the striatal sector receiving medial PFC afferents) and the dorsolateral (sensorimotor) striatum, which does not receive projections from the medial PFC. When analyzed in this fashion, PFC DA depletion did not augment TH activity or DA utilization in the striatal complex (Rosin et al., 1991).

Effects of prefrontal cortical DA depletions of the responsiveness of subcortical DA systems to environmental and pharmacological challenges

We have been unable to replicate the findings reported by Pycock, Carter, and associates, namely that PFC DA depletion enhances subcortical DA utilization. A number of other investigators have also been unable to demonstrate increased subcortical DA function following cortical 6-OHDA lesions, while still other investigators have been able to replicate certain aspects of the original studies (for a discussion of this issue, see Rosin et al., 1992). What factors may account for such discrepant results?

Certain variables that may contribute to the different findings do not appear critical. Thus, the sex, age, and strain (but not source) of rats used appear to be comparable in at least some of the studies reporting divergent results. Similarly, the time between surgery and sacrifice is comparable in a number of studies, such that evidence of increased DA utilization has been reported at both short (5–7 days) and long (≥30 days) post-operative intervals, whereas negative results have also been reported at times ranging from 14 to 30 days or more after surgery.

Another factor that may be relevant is the degree of sparing of the PFC NE innervation following 6-OHDA lesions. Carter and Pycock (1980) reported that subcortical DA function was unaltered in animals injected with 6-OHDA but not pretreated with the noradrenergic reuptake blocker DMI. This observation suggests that some degree of sparing of noradrenergic fibers is necessary in order for cortical dopamine lesions to regulate subcortical function. Regulatory controls over corticofugal projection neurons involve contributions from, and interactions of, DA and NE (Bunney and Aghajanian, 1976; Chiodo and Bunney, 1984; Tassin et al., 1986; see also Tassin et al., this volume). DMI pretreatment, however, has been used both in studies reporting positive as well as negative findings. With a single exception (Leccesse and Lyness, 1987), all investigators have reported that pretreatment with DMI attenuates but does not prevent NE depletion. The ratio of noradrenergic to dopaminergic involvement in the lesion area may represent an important variable.

Another potential confounding variable that has not been addressed is the degree to which animals are stressed immediately prior to sacrifice. For example, moving animals from the vivarium to the quarters in which the animals are sacrificed may have important consequences. Mild stressors have been known for some time to result in the metabolic activation of the PFC DA system (Thierry et al., 1976; Deutch et al., 1985), while somewhat more intense stressors result in the concurrent activation of the PFC and NAS DA innervations (see Deutch and Roth, 1990). Thus, a simple transfer of animals the night before or the day of the experiment may serve as a mild stressor. Similarly, the smell of stressed conspecifics or blood may serve as a stressor; this possibility requires that care be devoted to the manner in which animals are transferred at the time of sacrifice, and the organization of animals quarters in relation to experimental rooms. If an animal is mildly stressed prior to sacrifice, we reasoned that DA release in the nucleus accumbens might be somewhat enhanced, but not to a sufficient degree to be reflected in a statistically significant manner. When such animals are subjected to mild stressors that result in the overt metabolic activation of the PFC but not of the NAS, the resultant combination of sub-threshold increases in DA release in the NAS by stress and cortical DA depletion may summate to result in an overt increase in mesolimbic DA release. We therefore examined this possibility by subjecting animals with prefrontal cortical DA depletion to mild footshock stress (Deutch et al., 1990).

Exposure to footshock stress for 20 minutes increased DA utilization in the PFC but not in the NAS of control (vehicle-infused) animals. Dopamine metabolite levels and DA utilization were not increased in the NAS of rats which received 6-OHDA lesions of the PFC but were not stressed. However, in those lesioned animals subsequently exposed to footshock, there was a statistically significant increase in DA utilization in the NAS (Deutch et al., 1990). Thus, although PFC DA depletion did not increase *basal* mesolimbic DA function, the lesion enhanced the *responsiveness* of the NAS DA innervation to stress.

The enhanced responsiveness of mesolimbic DA neurons to an environmental challenge may reflect an intrinsic adaptive change in these neurons, which renders them more susceptible to some extrinsic challenge. In order to examine this possibility, we examined the responsiveness of subcortical DA neurons to a low-dose haloperidol challenge in animals with PFC DA depletions; the increase in DA synthesis and release following acute low-dose haloperidol administration predominantly occurs in response to autoreceptor blockade (Braszko et al., 1981; Westerink and de Vries, 1989). Lesioned and vehicle-infused controls received 0.05 mg/kg haloperidol; this dose of the neuroleptic results in a sub-maximal increase in TH activity in the dorsal and ventral striatum. Lesioned animals did not exhibit any increase in TH activity relative to vehicle-infused controls in the subcortical sites examined (NAS, medial periventricular striatum, dorsolateral striatum, substantia nigra, ventral tegmental area). However, the haloperidol challenge increased NAS TH activity in the lesioned animals to a significantly greater degree than in control subjects (Rosin et al., 1991). Preliminary data from our laboratory suggest that haloperidol also increases DA release in the NAS, as assessed by *in vivo* microdialysis, to a significantly greater degree (but for a similar duration) in animals with cortical DA lesions. These data therefore indicate that the DA innervation of the NAS in animals with PFC DA depletions is significantly more responsive to both environmental (stress) and pharmacological (haloperidol) challenges.

A key question that develops from these studies is why basal DA utilization is not increased, yet under conditions of challenge the DA system in the NAS is hyperresponsive. This situation may arise because the lesions of the PFC DA innervation are incomplete. In order to effect a complete loss of DA, the lesion parameters have to be changed such that there is concomitant damage to non-catecholaminergic elements in the PFC. The residual DA in the PFC (which is typically on the order of 20% of control levels) may be sufficient to inhibit corticostriatal neurons under basal conditions, and thus result in no net change in glutamate efflux in the subcortical sites. However, under conditions of perturbation, such as mild stress or haloperidol, the DA loss becomes apparent because of increased demand. Thus, in the case of haloperidol challenge, any post-synaptic blockade is added onto a very weak ability of acute haloperidol to evoke PFC DA release (Moghaddam and Bunney, 1990a). In the case of lesions of the PFC DA innervation augmenting the responsiveness of the NAS to stress, one must posit that stress — in addition to enhancing cortical DA release — directly drives PFC neurons. We have observed that neurons in the PFC express the immediate-early gene c-*fos* in response to stress, suggesting that these neurons may be activated (see below); the source of a stress-elicited excitatory drive to PFC neurons is not known. If the PFC DA response to stress is part of a coping response (see Deutch and Roth, 1990; D'Angio et al., 1990), then the normal ability of DA to inhibit a stress-evoked increase in corticofugal neuronal activity would be severely compromised by the lesion, and an increase in subcortical DA tone would result.

Mechanisms of prefrontal cortical regulation of subcortical DA systems

The enhanced responsiveness of subcortical DA systems to various challenges following lesions disrupting the dopaminergic innervation of the PFC may help explain the difficulties a number of laboratories have encountered in attempts to replicate the original reports of Carter and Pycock. Under the conditions of these experiments, the DA innervation of the PFC was lesioned and animals were allowed to recover for a variable period of time prior to sacrifice. During this post-operative interval, a number of adaptive changes may occur, both in the DA neurons innervating the subcortical sites and in non-dopaminergic afferents and intrinsic elements of the striatal complex; among such adaptive responses may be changes in receptor number or affinity, or alterations in TH gene expression or protein phosphorylation. While such adaptive changes are likely to reflect parallel adaptive changes encountered in human disease states, including schizophrenia, it is useful to obtain data addressing the issue of PFC DA regulation of subcortical DA systems in animals without lesions, thus obviating the possibility of adaptive changes. Studies using *in vivo* voltammetric methods have led to the conclusion that the subcortical DA systems are indeed transsynaptically regulated by corticostriatal neurons under tonic DA inhibition. Thus, local PFC infusion of the DA receptor antagonists cis-flupenthixol or sulpiride enhance indices of DA release from the NAS, whereas infusions of the indirect DA agonist amphetamine decrease DA release in the mesolimbic site (Louilot et al., 1987, 1989). More recently, Jaskiw et al. (1991) have used in vivo microdialysis to demonstrate that infusions of the DA agonist apomorphine into the PFC reduce extracellular DA levels in the striatum.

While it has become clear that the mesoprefrontal cortical DA system transsynaptically regulates subcortical DA systems, answers to a number of important questions remain elusive. Among these are the precise mechanisms through which changes in corticostriatal function may impact on subcortical DA release, and the precise sites at which such transsynaptic regulation of subcortical DA function is occurring. It is paradoxical to note that although the phenomenology has now been relatively well established, the mechanisms through which cortical DA system can regulate subcortical sites have become less clear. In particular, the nature of glutamate-induced DA release has become more complex than hitherto realized.

Glutamatergic regulation of striatal DA release

Although many studies have indicated that glutamate can evoke striatal DA release via presynaptic (impulse-independent) means, most studies have been able to evoke striatal DA only by using very high concentrations of glutamate. A key methodological variable may have contributed to the different findings: most of the positive findings have been obtained in studies examining the release of [^{3}H]-DA, synthesized from exogenous [^{3}H]-tyrosine (Giorguieff et al., 1977; Roberts and Anderson, 1979; Cheramy et al., 1984, 1986), whereas the negative findings have typically examined endogenous release of DA (Jhamandas and Marien, 1987; Chow and Jhamandas, 1989). The ability to observe glutamate-evoked DA release in those studies using a radiolabelling approach may, in part, derive from the fact that glutamate appears to increase TH activity, and thus there would be a greater amount of the newly synthesized (preferentially released) pool of [^{3}H]-DA available from the labelled precursor. Consistent with this suggestion is the observation that high concentrations of the amino acid are also required to evoke the release of [^{3}H]-DA after preloading with [^{3}H]-DA (Marien et al., 1983).

Recent studies of the ability of glutamate and other EAA to regulate DA release have employed *in vivo* microdialysis to assess endogenous DA release in response to local infusion of glutamate or other EAA receptor agonists (Bennett and Leslie, 1990; Moghaddam et al., 1990a; Payson and Donzanti, 1990). Such studies have typically observed that very high concentrations of glutamate or related EAA are required to increase extracellular concentrations of DA. The fact that such high concentrations of glutamate are required have led some investigators to speculate that glutamate-evoked DA release may be not be physiologically relevant. This concern is further amplified by the observation that EAA receptor antagonists do not appear to alter basal DA efflux (Keefe et al., 1990; Moghaddam and Gruen, 1991). Moreover, the paucity of axoaxonic synapses between corticostriatal axons and striatal dopaminergic afferents has bolstered the concept that glutamate-evoked DA release is a non-specific effect of the amino acid and is not of physiological relevance.

Electron microscopic studies concerning the precise localization of EAA receptors on various striatal elements has not been performed. However, it appears likely that an AMPA family (non-NMDA) receptor is present on DA afferents to striatum, since we have recently localized an AMPA family receptor mRNA to neurons of the substantia nigra, pars compacta (Wong and Deutch, 1991).The presence of such a receptor on DA neurons is consistent with the suggestion that a uniform kainate/quisqualate type receptor mediates impulse-independent DA release evoked by glutamate (Barbieto et al., 1991), whereas a NMDA type receptor localized to striatal neurons (Errami and Nieoullon, 1986; Samuel et al., 1990) mediates impulse-dependent, TTX-sensitive glutamate-evoked DA release (see also Krebs et al., 1991).

The lack of conventional synapses between dopaminergic and cortical axons in striatal complex cannot be interpreted to indicate that there is no morphological substrate for glutamate-evoked DA release. Over the past decade it has become increasing clear that so-called "mismatches" between transmitter inputs and receptors are a common occurrence in the CNS (see Herkenham, 1991). A recent series of studies by Beaudet and colleagues are instructive. These investigators have elegantly demonstrated a mismatch between neurotensin (NT)-containing afferents to the DA neurons of the ventral mesencephalon and NT receptors located on these DA neurons. NT axons make circumscribed synapses onto DA neurons in the substantia nigra. However, the localization of NT receptors are often at other sites on the neuron, and indeed are distributed along the entire neuronal membrane. Thus, sites of synaptic contact between NT afferents and DA neurons do not necessarily correspond to sites of NT receptor localization. It has been suggested that there may be two distinct types of regulation by NT of DA neurons: a direct excitatory effect, mediated by the NT synapses which on the post-synaptic density contain NT receptors, and a second, paracrine mode of modulatory regulation over DA neurons at sites which are not congruent for NT synapses and receptor localization (see Beaudet and Wolfe, 1992; Deutch and Zahm, 1991). The latter mode of regulation would require diffusion of the released NT in order for the transmitter to exert a modulatory role over DA transmission. In turn, it appears likely that local concentrations of the transmitter would have to be quite high in order to escape inactivation by peptidases. Following this reasoning, it is not unreasonable to posit a similar requirement for relatively high glutamate concentrations to evoke DA release from striatal DA axons in the absence of conventional synaptic arrangements.

Glutamate also differs from most neurotransmitters in that it not only serves as a transmitter, but also as a metabolic intermediary. The latter function of glutamate and related amino acids is associated with the uptake of glutamate by glia. The avidity of the glial glutamate sink is such that much higher concentrations of the amino acid are required to achieve sufficient extracellular glutamate concentrations to function as a neurotransmitter.

The requirement for what appears to be unusually high glutamate concentrations may also be related to the characteristics of the receptor that mediates EAA actions. As noted earlier, DA neurons express an AMPA receptor gene. The uniform kainate/quisqualate site undergoes very rapid desensitization (Mayer and Vyklickly, 1989), and thus high concentrations of the ligand may be required to evoke DA release. Supporting this contention is the observation of Barbieto et al. (1991) that striatal infusion of concanavalin A, which prevents desensitization of the receptor (Mayer and Vyklickly, 1989), markedly increases the amount of DA released in response to glutamate. The striatal NMDA receptor also undergoes desensitization upon exposure to agonists, but at a slower rate than that of the non-NMDA receptors (Lehmann et al., 1988).

The requirement for high concentrations of EAA may in part be related to which amino acid serves as the endogenous ligand. Our knowledge of the endogenous ligands for various EAA receptors is in its infancy. In the absence of precise knowledge of the endogenous transmitter, glutamate has been most often used. However, lower concentrations of other EAA, such as kainate or quisqualate, are required to evoke DA release than glutamate (Imperatu et al., 1991; Payson and Donzanti, 1990). The precise receptor localized to DA axons and its endogenous ligand must be defined before it will be possible to state the minimal concentration of the ligand required to evoke DA release. Moreover, without this knowledge it will be difficult to conduct appropriate antagonist studies.

Two final caveats need to be made concerning the apparently high concentration of glutamate required to evoke DA release. The first is that almost all studies to date of glutamate-dopamine interactions have been performed in the dorsal striatum. We have been able to observe effects of PFC DA depletion on subcortical DA function only in the NAS, but not in the dorsal striatum. The organization of neuronal elements in the dorsal and ventral striatum differ; similarly, transmitter interactions in the dorsal and ventral striatum differ (Consolo et al., 1977; Girault et al., 1986a,b). There may be corresponding differences in the degree to which EAA evoke DA release from different dopaminergic terminal fields (Marien et al., 1983; Payson and Donzanti, 1990; Imperatu et al., 1991). Future studies focusing on both the ventral striatum (including both the NAS and olfactory tubercles) and dorsal striatum will be required to unravel the relative efficacy of EAA in the regulation of DA systems in the striatal complex.

A second caveat concerns the methodology employed in the recent studies of endogenous DA release by glutamate. *In vivo* microdialysis offers many advantages for the study of transmitter release, but, as with all methods, there are also attendant disadvantages. The use of *in vivo* microdialysis to study DA release may modify autoregulatory tone over DA neurons by removing extracellular DA. Such a decrease in autoinhibition would increase DA release beyond normal baseline levels to an artificially elevated state. Even relatively small increases in DA release above true baseline control levels would make it more difficult to observe an enhancement of DA release by glutamate or related amino acids. Push-pull cannulation techniques are plagued by the same constraint. Finally, the magnitude of change in extracellular DA concentrations as measured by dialysis is often discordant with the magnitude of change in both *ex vivo* neurochemical markers and behavior (see Kalivas and Duffy, 1990); the most parsimonious explanation for this discrepancy may simply be that *in vivo* dialysis underestimates evoked DA release. While both *in vivo* dialysis and push-pull cannula methods offer a unique opportunity to examine the dynamics of DA release in the non-anesthetized, freely-moving animal, suitable caution must be exercised when interpreting the results of such studies.

Further investigations will be required to resolve the discrepancies concerning such issues as the concentrations of amino acids required to evoke striatal DA release. Virtually all studies of the EAA regulation of DA release have been performed using either *in vitro* approaches to the study of striatal tissue, or alternatively *in vivo* methods to assess striatal DA release. There have been no direct studies of the involvement of glutamate in the ability of DA afferents to the cortex to transsynaptically regulate subcortical DA release. Lesion studies of the cortex do indeed suggest that ablation of the cortex or transection of the corticostriatal projections decrease concentrations of glutamate and aspartate, but there have been very few studies of the effects of cortical stimulation on striatal EAA release. Such studies are clearly possible; for example, Palmer et al. (1989) reported that local administration of the GABA antagonist bicuculline to the cortex increased extracellular concentrations of aspartate but had no effect on glutamate levels. There have been no comparable studies involving direct pharmacological manipulation of the cortical DA system to examine the effects of DA and EAA release in the subcortical DA territories. Moreover, it is necessary to determine to what degree EAA released from corticostriatal terminals exert a tonic, as opposed to phasic, regulation of striatal DA release. Such studies involving concomitant measurement of DA and EAA will be necessary before we can understand the nature of glutamate-evoked DA release and its functional determinants.

Multiple cortical dopamine systems and the regulation of subcortical dopamine function

The interpretation of our observations that PFC DA lesions alter subcortical DA function is predicated on the ability of EAA from disinhibited corticostriatal axons to presynaptically evoke DA release. While there is a large body of data supporting this conceptualization, other data are discordant. An understanding of the physiologically relevant bases for the ability of cortical DA systems to transsynaptically regulate striatal DA release will require increased attention to the physiological and behavioral context of the studies.

As noted previously, the stress-evoked increase in mesolimbic DA release is augmented under conditions of prefrontal cortical DA depletion (Deutch et al., 1990). We have interpreted this observation, and our subsequent finding that ibotenic acid lesions of the PFC suppress stress-evoked DA release from the NAS, to indicate that the transsynaptic regulation of mesolimbic DA axons by the PFC governs, in part, the responsiveness of the NAS DA innervation to stress.

Exposure to mild stressors results in a selective increase in DA metabolism and release in the PFC (Thierry et al., 1976; Deutch et al., 1985); increases in either the duration or intensity of the stressor enhance DA release in the NAS, and ultimately in the dorsal striatum (Roth et al., 1988;

Deutch and Roth, 1991). We have speculated that the release of DA from the subcortical DA innervations, such as the NAS, is in part independent of increases in the firing rate of mesolimbic and nigrostriatal DA neurons, while the change in PFC DA release is attributable to a stress-induced increase in the firing rate of mesoprefrontal cortical DA neurons (Deutch and Roth, 1990). Our speculation was prompted by electrophysiological data obtained by Thierry, Mantz, and colleagues (Mantz et al., 1989; Thierry et al., 1990). These investigators reported that noxious stimulation of anesthetized rats increased the firing rate of identified mesoprefrontal cortical DA neurons. In contrast, noxious tail pinch did not increase the discharge frequency of A10 DA neurons innervating the nucleus accumbens or septum. Moreover, non-noxious tactile stimulation did not enhance the firing rate of mesoprefrontal cortical neurons.

We have recently described a similar differentiation of the responsiveness of mesoprefrontal and mesoaccumbens DA neurons in the rat. The immediate-early gene c-*fos* is thought to be a key transduction mechanism regulating gene expression of a number of neurotransmitters (Sheng and Greenberg, 1990). It was originally suggested that the presence of the protein product of c-*fos*, termed Fos, served as a sensitive index of neurons which are physiologically activated (Sagar et al., 1987). It has recently become clear that this is the case only for certain neurons. However, it does appear that in those neurons capable of expressing Fos the presence of the protein is a relatively sensitive index of neuronal metabolic activation. It should be noted that such metabolic activation does not necessarily correspond to depolarization, and can occur in the presence of intracellular changes that accompany modulatory influences (Morgan and Curran, 1986). With these caveats in mind, we examined the stress-induced changes in Fos expression in midbrain DA neurons (Deutch et al., 1991b).

Exposure of rats to 30 minutes of restraint stress increased DA metabolism not only in the PFC but also in the NAS and dorsal striatum; 15 minutes exposure to the stressor increased DA metabolism in both the limbic and cortical sites, but not dorsal striatum. Under these stress conditions, a dramatic increase in the number of ventral tegmental area (A10) DA neurons expressing Fos protein was observed; in contrast, Fos expression did not increase in A9 (substantia nigra) or A8 (retrorubral field) DA neurons. Moreover, the anxiogenic benzodiazepine inverse agonist FG 7142 also increased the number of A10 (but not A8 or A9) DA neurons expressing Fos. The fact that only VTA DA neurons expressed Fos protein after stress suggested that the increase might be restricted to a distinct subpopulation of mesotelencephalic DA neurons, since A8 and A9 DA neurons project to mesolimbic and striatal sites (Deutch et al., 1988). Further examination revealed that the increase in the number of DA neurons expressing Fos after stress was heterogeneous within the VTA, mainly being localized to the nucleus parabrachialis pigmentosus and caudal linear nucleus. These two regions of the VTA provide part of the projection to the PFC (Swanson, 1982). We therefore examined Fos expression in DA neurons retrogradely labelled from the PFC or NAS. Certain (but not all) A10 DA neurons projecting to the PFC increased expression of Fos in response to stress; in contrast, not a single DA neuron that was retrogradely labelled from the NAS expressed Fos.

While these data are consistent with the speculation that stress results in an increase in the firing rate of mesoprefrontal cortical DA neurons, but not mesoaccumbens DA neurons, the difficulties in correlating Fos expression and depolarization preclude a more definitive statement. More interesting, perhaps, is the fact that only a distinct subpopulation of DA neurons retrogradely labelled from the PFC expressed Fos protein after stress. These data therefore indicate that the dopaminergic innervation of the PFC is heterogenous, and is not a single functional entity. Thus, there are at least two populations of mesoprefrontal cortical DA neurons: those that express Fos after exposure to stress and those that do not.

The efferent projections of the medial PFC are topographically organized, both across and within the different cytoarchitectonic areas that comprise the region (Sesack et al., 1989; McGeorge and Faull, 1989); on these grounds and the available physiological data we have tentatively ascribed functions to the different parts of the PFC (Sesack et al., 1989). It is therefore not unreasonable to assume that the DA innervations of different parts of the rodent PFC are involved to varying degrees under different physiological demands. For example, one might expect that mild footshock stress would result in an increase in DA utilization in the ventral prelimbic (area 32) and infralimbic (area 25) cortices, the area of the PFC which may be critical for autonomic arousal. We therefore examined the biochemical heterogeneity of the dopaminergic innervation of the medial prefrontal cortex (Deutch, 1991).

Animals were subjected to mild footshock stress, sacrificed, and the different regions of the medial PFC dissected. Concentrations of DA and its metabolites, serotonin and its primary acidic metabolite, and NE were measured. Mild stress did indeed have a differential effect on DA utilization in the PFC. Specifically, the rostral PFC responded to stress by increasing DA utilization both in the shoulder cortex (rodent homologue of the premotor cortex, supplementary motor area, and frontal eye field [Passingham et al., 1988]) and in the rostral prelimbic cortex. In contrast, the DA innervations of the caudal prelimbic and shoulder cortices did not respond to the stressor, but there was a highly significant increase in DA utilization in the infralimbic cortex. Finally, the stress-induced biochemical activation of the PFC was restricted to the deep layers of the cortex; there was no change in DA utilization in the PFC superficial layers after stress exposure. Lamina-specific differences in the function of PFC DA systems have been proposed (Berger and Verney, 1984; Berger et al., 1991; see Lewis, this volume).

The heterogeneity of the PFC DA response to stress is consistent with our earlier data indicating that stress could be used to define at least two functionally distinct groups of mesoprefrontal cortical DA neurons. The Fos data may also suggest that there are distinct attributes of different mesoprefrontal cortical DA neurons at the somatodendritic level which dictate the response characteristics we have observed in the terminal fields; the topographical distribution of distinct afferents to the VTA may be such an attribute. However, it is not clear to what degree mesoprefrontal cortical DA neurons arborize to distribute over a wide spatial domain. If a single DA neuron gives rise to an extensive terminal network, spreading beyond the confines of a single architectonic region, the differences in response to stress in the different cytoarchitectonic areas comprising the PFC are presumably related to presynaptic regulation of DA release. We have previously demonstrated such presynaptic regulation of DA metabolism in the PFC: frontal cortical tissue slices incubated in the presence of the anxiogenic β-carboline FG 7142 increase tyrosine hydroxylation, whereas PFC slices incubated with γ-aminobutyric acid (GABA) decrease DA synthesis (Knorr et al., 1989). We are currently performing anatomical studies to determine to what degree the DA innervations of the different parts of the PFC arise from different populations of VTA neurons.

The fact that different PFC sectors innervate different subcortical regions suggests that the difficulty in observing functional consequences of PFC DA depletions on basal subcortical DA systems may be attributable, in part, to lesions that do not involve the appropriate part(s) of the PFC. While an inability to observe an effect of PFC DA depletion on basal DA function in subcortical sites may reflect the incomplete nature of the lesion (in a fashion analogous to that observed in the striatal system after partial lesions of the nigrostriatal DA system [Hefti et al., 1980; Zigmond et al., 1990]), it is perhaps more likely that lesions in different studies examining the effects of PFC DA depletion on subcortical DA function impact to varying degrees on dopaminergic axons in different parts of the PFC, with some sparing of certain PFC fields. Only one published investigation has provided a full histological analysis of the lesion area (Rosin et al., 1991). Based on the stereotaxic coordinates provided, several studies appear to have directed injection cannulae at more dorsal sites within the PFC, and thus probably spared the ventral prelimbic and infralimbic cortices. Although our lesions did involve the prelimbic and dorsal infralimbic cortices (Rosin et al., 1991), which are the same areas in which DA axons are impacted by stress, we failed to observe an increase in basal DA utilization in subcortical targets to which these sites project heavily (e.g., the nucleus accumbens). It therefore seems likely that an interaction of site and animal state are important in determining the effect of PFC DA depletion on subcortical DA systems.

Differential responsiveness of the dorsal and ventral striatum to manipulations of the PFC DA system

Our investigations of the effects of PFC DA depletion on subcortical DA systems have revealed an increased responsiveness of mesolimbic, but not nigrostriatal, DA neurons to both environmental and pharmacological challenges. These data thus call into question the site(s) at which corticostriatal neurons impact on subcortical DA systems.

Manipulations of cortical DA systems can affect subcortical DA systems at either the somato-dendritic (midbrain) or terminal field (dorsal and ventral striatum) levels. Projections from the PFC to the midbrain have been well documented (Beckstead, 1978; Sesack et al., 1989) and include EAA projections to both the ventral tegmental area and medial substantia nigra (Carter, 1982; Christie et al., 1985). In our anatomical studies, we have not observed a projection from the PFC to the retrorubral A8 DA neurons (Deutch et al., 1988; Sesack et al., 1989), although fibers of passage perforate the region.

The projections from the cortex to the midbrain provide an avenue whereby alterations in PFC DA systems may impact on subcortical (mesolimbic and striatal) terminal fields. Such changes would presumably occur via changes in either the firing frequency or firing pattern (Gariano and Groves, 1989) of midbrain DA neurons. If PFC DA depletions alter DA utilization in the striatal complex through changes occurring at the cell body level, the interpretation of presynaptic glutamate-evoked DA release becomes moot. However, we have not observed a change in any index of DA function (*in vitro* TH activity, *in vivo* tyrosine hydroxylation, DA utilization), under either basal or stimulated conditions, in animals with PFC DA depletions. Moreover, even those studies reporting a change in striatal DA function following PFC DA depletions do not report a concurrent change in the midbrain cell body regions (Carter and Pycock, 1980; Pycock et al., 1980a,b). It therefore seems most likely that the effects of PFC DA depletion of the striatal DA systems are not manifested through changes in the firing rate (or pattern of firing) of midbrain DA neurons, but probably represent a presynaptic regulation of striatal DA release.

If the effect of PFC DA lesions is manifested at the terminal field level, how does one explain the consistent lack of change in DA function in the dorsal striatum, while the ventral striatal (NAS) DA system is susceptible to PFC manipulations? Two mechanisms come to mind: different synaptic relationships may be operative in the dorsal and ventral striatum, or the change in the dorsal striatum is restricted to certain macro- or micro-scopic compartments in the dorsal striatum.

The dorsal and ventral striatum can be differentiated on a number of functional grounds, including differences in transmitter interactions and differences in behavior subserved. Virtually all studies of the effects of EAA on DA release have examined the dorsal striatum. The TTX-sensitive release of DA by EAA appears to occur through an NMDA receptor present on striatal cholinergic cells (Samuel et al., 1990), where L-homocysteic acid has been advanced as the endogenous ligand (Cox et al., 1977; Do et al., 1986; Knopfel et al., 1987; Lehmann et al., 1988). However, such a cholinergic mediation of DA release (evoked by glutamate stimulating an NMDA receptor) would presumably be operative only in the dorsal, but not ventral, striatum, since the NAS appears to lack a functional cholinergic-dopaminergic link (Consolo et al., 1977). At present it is not possible to state the degree to which TTX-sensitive glutamate-evoked DA release may alter the capacity of the endogenous ligand for the non-NMDA site to evoke (TTX-insensitive) DA release. However, it does seem clear that direct examination of glutamate-dopamine dynamics in the ventral as well as the dorsal striatum (see Payson and Donzanti, 1990; Imperaturu et al., 1991) represents an important area for future investigation, as does the relative involvement of cholinergic and GABAergic neurons.

The second possible explanation for the lack of effect of PFC DA depletion on basal subcortical DA function may be the restriction of changes in DA dynamics to distinct striatal compartments. For example, the examination of the dorsolateral striatum for changes in DA function after PFC lesions would appear to be a futile exercise, since the medial PFC does not project to this striatal sector. It was for this reason that we examined a PFC-recipient zone of the dorsal striatum (medial periventricular striatum) as well as a non-recipient zone (dorsolateral striatum) as a control site; PFC DA lesions did not alter DA function in either striatal region.

A second level of striatal compartmental organization is reflected by the localization of various histochemical markers to distinct territories termed patch (or striosome) and matrix. There are two distinct DA innervations of the dorsal striatum that can be differentiated on developmental, hodological, and pharmacological grounds (see Gerfen, this volume). The islandic DA innervation appears early in development (Olson et al., 1972), arises from a spatially distinct population of nigral DA neurons (Gerfen et al., 1987), is in register with the cholinesterase-poor (striosomal) compartment of the striatum (Graybiel, 1984), and has different responses to certain ergolene DA agonists (Fuxe et al., 1978). The more widespread diffuse (matrix) DA innervation of the dorsal striatum appears during postnatal development and is in register with the cholinesterase-rich (matrix) compartment.

The medial PFC was originally suggested to selectively innervate the striatal patch compartment, whereas other cortical areas (such as somatosensory cortex) projected to the matrix (Donoghue and Herkenham, 1986). A reevaluation of the projections from the medial PFC has revealed that the PFC neurons in the superficial aspects of layer V project to the matrix, whereas deep layer V and VI PFC neurons project to the patch compartment (Gerfen, 1989; see Gerfen, this volume). Thus, it is possible that the effects of 6-OHDA-induced lesions of the PFC on striatal DA systems may be predominantly localized to the DA innervation of the patch compartment, i.e., the islandic DA innervation. There have been no direct examinations of the effects of PFC DA depletions on the islandic and diffuse striatal DA systems. If PFC DA depletions do indeed alter DA dynamics in the dorsal striatum, a selective effect on the islandic DA innervation (which comprise approximately 15% volume of the striatal DA system) would be extremely difficult to discern using conventional methodologies. Alternative approaches, including quantitative microspectrophotofluorometric assessment of DA turnover (Agnati et al., 1979) or the use of a micro-push-pull cannulation system, may allow an assessment of the relative involvement of the islandic and diffuse DA innervations after PFC DA lesions.

Elaboration of parallel corticostriatal dopamine systems

The finding that the DA innervation of the PFC is not functionally homogeneous is consistent with previous reports indicating that there are developmental gradients in the dopaminergic innervation of the frontal cortices (Berger and Verney, 1984; Berger et al., 1991), and that different populations of mesoprefrontal cortical DA neurons can be identified on the basis of the presence of impulse-modulating autoreceptors (Shepard and German, 1984). Moreover, we have shown that there is a differential vulnerability of anterior cingulate DA innervations in the primate (which best corresponds to area 24b in the rat medial PFC) to the neurotoxic actions of 1-methyl-4-phenyl-1,2,2,6-tetrahydropyridine (MPTP; Elsworth et al., 1990).

Recent physiological and anatomical data have led to a theory of parallel but segregated cortico-striato-pallidal loops through which information is processed, such that each loop may subserve a particular function (Alexander et al., 1986; Alexander and Crutcher, 1990). The physiological data that led to this formulation were predominantly derived from studies of motor control in primates.

We have indicated that dorsal striatal dopaminergic systems may be separated into the islandic and diffuse innervations, and raised the possibility that these two DA systems are differentially regulated by the PFC. Recent anatomical data indicate that the nucleus accumbens can be parcellated into two distinct compartments, designated the core and shell. These two compartments have different efferent projections. The NAS_{core} projects to the dorsolateral ventral pallidum, which in turn innervates the subthalamic nucleus and substantia nigra; the NAS_{shell} projects to the ventromedial aspects of the ventral pallidum, from where fibers to the ventral tegmental area emanate (Zahm and Heimer, 1990; Heimer et al., 1991). In addition, both accumbal sectors have direct projections to the midbrain, with the NAS_{core} more heavily investing the lateral aspects of the ventral mesencephalon (substantia nigra), whereas the NAS_{shell} projects to the more medially situated VTA. These anatomical data have led to the speculation that the NAS_{shell} may be a mesolimbic component, whereas the NAS_{core} more closely resembles a dorsal striatal sector; anatomical (Zahm, 1991) and pharmacological (Deutch and Cameron, 1991) data support this formulation.

We determined that although both the NAS_{core} and NAS_{shell} exhibit a marked increase in DA utilization in response to haloperidol and clozapine, the NAS_{shell} DA innervation is significantly less responsive to haloperidol challenge that the NAS_{core}, while the two sectors respond equivalently to the atypical antipsychotic drug clozapine (Deutch and Cameron, 1992). These data are consistent with the observation that neuroleptic drugs evoke more DA release from the dorsal striatum than NAS, while atypical antipsychotic drugs increase DA release to an equivalent degree in mesolimbic and nigrostriatal regions (Bartholini, 1976a,b; Julou et al., 1976). Moreover, restraint stress for 15 minutes selectively increased DA utilization in the NAS_{shell} but not in the NAS_{core}; the NAS DA innervation is significantly more susceptible to stress than the dorsal striatal DA innervation (Deutch and Roth, 1990).

The functional differentiation of the NAS into two distinct sectors with different efferent projections, coupled with the observation that the DA innervations of different cytoarchitectonic regions comprising the PFC respond differently to stress, suggests that it may be possible to define specific functional circuits involving DA. For example, it may be possible to define a dopaminergic "stress circuit", involving DA afferents to the ventral prelimbic and infralimbic cortices, which in turn project to the NAS_{shell}, where the DA innervation also responds to stress; the NAS in turn projects both indirectly and directly to the VTA DA neurons, which have been shown to respond to stress (Deutch et al., 1985). In contrast, there may be a "premotor circuit" involving the DA innervations of the dorsal prelimbic cortex, the NAS_{core}, and the DA neurons of the substantia nigra. We have recently initiated anatomical studies defining the afferent projections of core and shell accumbal sectors (Brog et al., 1991). Anterograde tracing data suggest that the infralimbic cortex projects rather heavily to the NAS_{shell}, but in contrast provides only a very small afference to the core of the accumbens. In more dorsal aspects of the medial PFC, including the ventral aspects of the prelimbic cortex, retrogradely-labelled cells are seen after injections into both the core and shell, but are intermingled, while in the dorsal prelimbic cortex core-labelled cells predominate. These anatomical observations fit well with a proposed "stress circuit" that is defined by a distinct PFC DA innervation in register with the DA innervation of the NAS_{shell}, which in turn projects to the VTA DA neurons.

It is important to note, however, that our anterograde data do yet not indicate a PFC (or any other) site that exclusively innervates either the core or shell of the NAS. While the infralimbic cortex and the region surrounding the anterior taenia tecta projects heavily onto the NAS_{shell}, a much smaller and less dense projection to the NAS_{core} is present. These data suggest that although it may be possible to define parallel corticostriatal DA loops, it is unlikely that these loops are completely segregated in function. For example, biological responsiveness to stress, which is of such great importance to the survival of animals, would presumably involve a number of distinct yet interactive components. For example, one component might involve vigilance and autonomic arousal, recruited by relatively mild stressors (perhaps on a continuum with novelty). More severe stressors would be expected to evoke a motor response for escape. The former would primarily be reflected by changes in DA function in a ventral PFC-NAS_{shell}-

VTA DA circuit, the latter via a dorsal prelimbic PFC-dorsal striatal-SN loop, but under conditions involving the latter circuit the autonomic stress loop would also be functional.

Implications of prefrontal cortical DA regulation of subcortical DA systems for schizophrenia

The original impetus for our studies of the regulation of subcortical DA systems by the PFC DA innervation was the hypothesis that such a mechanism is central to the pathophysiology of schizophrenia. Increased subcortical dopaminergic tone in the presence of a functional decrease in DA transmission in the PFC would help explain why D_2 receptor antagonists are of utility in the treatment of the positive symptoms of schizophrenia, but are largely ineffective in relieving negative symptoms (see Goldstein and Deutch, 1991).

We were unable to observe an increase in basal subcortical DA function following cortical DA depletion. However, the enhanced responsiveness of the mesolimbic DA innervations to stress and to haloperidol challenge have certain implications for schizophrenia. The stress-induced exacerbation of the psychotic process has been suggested on the basis of clinical observations and studies (Brown et al., 1972; Vaughn and Leff, 1976; Nicholson and Neufeld, 1989). This acute exacerbation is typically manifested in florid, positive symptoms. Our description of an enhanced responsiveness of the NAS DA innervation to stress after PFC DA lesions is consistent with the clinical observations. Moreover, if a functional impairment of cortical DA subserves negative symptoms, our findings would suggest that those patients with prominent negative symptoms might be more at risk for an acute psychotic break after stress. In particular, we speculate that such patients would be unusually prone to an acute psychotic episode following a variety of psychosocial stressors that would be tolerated by patients with the more typical balanced clinical presentation between positive and negative symptoms (Andreason et al., 1990).

We also observed an enhanced responsiveness of the nucleus accumbens DA innervation to low dose haloperidol challenge in animals sustaining PFC 6-OHDA lesions. These data lead to the speculation that patients with prominent negative symptoms patients might be more prone to extrapyramidal side effects than patients presenting with dominant positive symptomatology. Similarly, it is possible that patients with prominent negative symptoms will be more likely to have a poor therapeutic response to conventional neuroleptic therapy; although this possibility has been anecdotally noted, compelling data for this contention are lacking (see Carpenter, 1991). However, extrapyramidal side effects may be more frequent in such patients (and possibly at lower doses of the neuroleptics) because of cortical DA dysfunction in a region corresponding to the shoulder cortex in the rat, thus impacting on the dorsal striatum and the NAS_{core}, an accumbal sector which is functionally and hodologically related to the dorsal striatum. A further implication of such a finding would be that patients with prominent negative symptoms may be more effectively managed with lower doses of antipsychotic drugs. Finally, such observations may have some relevance to the actions of atypical antipsychotic drugs (see Deutch et al., 1991a). The atypical antipsychotic drug clozapine has been reported to release DA to a greater degree in the PFC than in subcortical sites, in contrast to neuroleptics (Moghaddam and Bunney, 1990a), and hence may be more effective in the treatment of negative symptoms and not induce extrapyramidal side effects. While these speculations may be empirically examined in patients, it will be important to define what constitutes prominent negative symptoms or the deficit state, and the relationship of these concepts to cortical DA depletion (see Carpenter, 1991; Tandon and Greden, 1991).

The course of schizophrenia is frequently marked by periods of acute psychosis interspersed with periods of relative recovery but without *restitutio ad integrum* (Bleuler, 1950). The "waxing and waning" course of the disease is characteristic. Our experimental work indicates that PFC DA depletion results in an enhanced responsiveness to both environmental and pharmacological challenges. Bleuler (1950) remarked that "... exacerbations may be internally determined, or by physical influences such as an alcoholic intoxication ... or by sudden shock-like psychic causes". If a cortical DA deficit is indeed present in schizophrenia, it may confer upon the patient increased susceptibility to various perturbations and result in the residual deficits during the intervening periods.

In addition to the functional implications that PFC regulation of subcortical DA function may hold for schizophrenia directly, a separate issue of considerable clinical importance is the significance for co-morbid conditions. Drug abuse is frequently is seen in the schizophrenic population. The response magnitude to amphetamine, cocaine, and other indirect DA agonists might be anticipated to be augmented in patients with cortical DA depletions. Moreover, the rate of sensitization to these agents may be accelerated. Preliminary data from our laboratory suggest that behavioral sensitization to low doses of cocaine is more marked in animals with PFC DA depletions.

Summary and conclusions

This chapter has focused on the regulation of subcortical DA systems by the PFC, and elaborated a conceptual framework for dopaminergic corticostriatal systems. The transsynaptic regulation of subcortical DA systems by the PFC dopaminergic innervations is now fairly well established, and is most obviously manifested under conditions of corticostriatal demand. Thus, although DA depletion in the PFC does not elicit changes in parameters of DA function under basal conditions, enhanced DA function (as reflected by changes in TH activity and DA utilization) in rats with 6-OHDA lesions is seen when animals are either pharmacologically (low dose haloperidol) or environmentally (footshock stress) challenged. The difficulty in observing changes in basal subcortical DA function may be due, in part, to the fact that the DA innervation of the PFC is not functionally or anatomically homogeneous, but consists of multiple dopaminergic systems embedded in the PFC. Moreover, preliminary anatomical data suggest that these DA systems may be in (transsynaptic) register with different subcortical DA systems, and may therefore define functional units. While there has been considerable success in defining the conditions under which the PFC DA innervations may transsynaptically regulate subcortical DA function, less clear are the mechanisms subserving these system interactions. Particularly troublesome is our lack of knowledge of the means whereby glutamate and other EAA can evoke the impulse-independent release of DA. However, efforts directed toward this problem continue, and a resolution of the issue seems imminent.

The discussion of the issues of corticostriatal regulation has focused on the prefrontal cortex and has not addressed other cortical sites. Recent post-mortem histological studies of schizophrenic brains have suggested pathological changes in a number of cortex sites, including the prefrontal cortex and temporal lobe structures. Moreover, functional impairments may be manifested without obvious morphological changes. Although the regulation of subcortical DA systems by the PFC has been emphasized, it is unlikely that the pathology of schizophrenia is restricted to a single site (see Czernansky et al., 1991). The entire cortical mantle projects in a topographically defined manner onto the striatum. It is not unreasonable to assume that changes in the function of DA in various cortical areas may impact on corresponding striatal target sectors. The dopaminergic innervation of the cortex is more elaborated in man and other primates than in the rodent. Varying degrees of DA dysfunction in different cortical areas may contribute to the different clinical picture presented by different schizophrenic patients: for example, different symptoms may be expected to arise as a consequence of a functional impairment in dopaminergic transmission in the dorsolateral convexity and in the entorhinal cortex.

The difficulties of extrapolating from rodent species (in which most basic research is performed) to primate species, including man, are well known. The impetus for the recent basic research in cortical regulation of subcortical DA systems is rather unusual in that it is directly derived from a clinical hypothesis. It is therefore prudent to ask whether the morphological pathology observed in the schizophrenias is mimicked by PFC 6-OHDA lesions in rats. Recent pathological findings of decreased laminar thickness in certain cortical areas suggests that there may be an overt structural deficit in the schizophrenic patient (see Lewis, this volume). The decrease in thickness may arise, however, from a decrease in the number of neurons, or alternatively from a loss of afferents to the region. A loss of dopaminergic afferents to cortical sites might be an example of the latter case. An actual decrease in the number of neurons must be further defined in order to be meaningful: are pyramidal (projection) neurons or the smaller interneurons lost? Recent analyses of post-mortem brains from schizophrenic subjects indicates a selective loss of GABAergic interneurons in the prefrontal and anterior cingulate cortices (Benes et al., 1991), but no decrease in the number of pyramidal neurons. The decreased number of GABA neurons is paralleled by an increase in $GABA_A$ receptor density (Benes et al., 1992). Such an observation suggests that corticostriatal neurons may be removed from an inhibitory constraint. It is clearly more difficult to demonstrate a subtle loss of DA afferents to cortical sites, particularly in man, where cortical catecholaminergic markers are degraded very rapidly in the post-mortem interval (Black and Green, 1975; Gutman et al., 1987; Goldstein et al., 1988). It is tempting to speculate, however, that removal of two inhibitory influences over pyramidal neurons, namely dopaminergic afferents and GABAergic interneurons, may have rather profound consequences on corticostriatal function, and thence on DA dynamics in subcortical sites.

The definition of the specific site(s) at which the pathology of schizophrenia is manifested will be of exceeding importance. For example, Grace (1990) has recently published a provocative hypothesis of the pathophysiology of schizophrenia (see also Grace, this volume). The endpoints of his hypothesis concerning dopaminergic alterations match quite closely some of the arguments concerning subcortical DA tone which I have presented, yet we are not in complete agreement concerning the etiology of these changes. Different viewpoints concerning the initial site of dysfunction become of considerable importance in attempts to develop new treatment strategies for schizophrenia. Such strategies require an awareness not only of the ultimate expression of the defect in DA systems, but in addition the definition of any antecedent neural systems which are dysfunctional and which may determine a final and common alteration in DA systems downstream. Targeting both initial defects and defining convergent paths onto common DA systems may enable the development of treatment strategies that operate through different transmitter systems, and may therefore generate specific therapeutic strategies for the different schizophrenic reactions.

In summary, it appears that the dopaminergic neurons comprising the mesotelencephalic DA system function interdependently. The interconnections at the terminal field level appear to be the primary determinant of integrated response characteristics, although interconnections directly at the cell body level (see Deutch et al., 1988) or indirectly via long-loop feedback projections from terminal fields to cell bodies may play an additional role. The interdependence of the mesotelencephalic DA neurons may offer important clues in the search for the pathogenesis of schizophrenia, and may have equally important implications for the clinical management of the disorder.

Acknowledgments

This manuscript is dedicated to Dr. L. J. Peacock on the occasion of his retirement. Prof. Peacock had the stamina to guide me through my graduate education, and without his guidance and wisdom I would not have succeeded. His devotion to science and experimental design, as well as the lessons to be learned from history, illuminated for me and other students a new world.

The insightful comments of Dr. E. M. Stricker markedly improved the manuscript, both in content and expression. I also appreciate helpful discussions with Dr. B. Moghaddam. This work was supported by grant MH-45124 and by support from the Schizophrenia Research Center and the Post-Traumatic Stress Disorder Center of the West Haven Veteran's Administration Medical Center, and the National Parkinson Foundation Research Laboratories at Yale University.

References

Agnati LF, Andersson K, Wiesel F, Fuxe K (1979) A method to determine dopamine levels and turnover rate in discrete nerve terminal systems by quantitative use of dopamine histofluorescence obtained by the Falck-Hillarp methodology. J Neurosci Methods 1: 365–373

Alexander GE, Crutcher MD (1990) Functional architecture of basal ganglia circuits: neural substrates of parallel processing. TINS 13: 266–271

Alexander GE, DeLong MR, Strick PL (1986) Parallel organization of functionally segregated circuits linking basal ganglia and cortex. Ann Rev Neurosci 9: 357–381

Andreason NC, Flaum M, Swayze VW, Tyrell G, Arndt S (1990) Positive and negative symptoms in schizophrenia: a critical reappraisal. Arch Gen Psychiatry 47: 615–621

Barbieto L, Cheramy A, Godeheu G, Desce JM, Glowinski J (1990) Glutamate receptors of a quisqualate-kainate subtype are involved in the presynaptic regulation of dopamine release in the cat caudate nucleus in vivo. Eur J Neurosci 2: 304–311

Bartholini G (1976a) Differential effect of neuroleptic drugs on dopamine turnover in the extrapyramidal and limbic system. J Pharm Pharmacol 28: 429–433

Bartholini G (1976b) Preferential effect of noncataleptogenic neuroleptics on mesocortical dopaminergic function. Adv Biochem Psychopharmacol 16: 607–611

Beaudet, Wolfe (1992) Morphological substrate for neurotensin-dopamine interactions in the rat midbrain tegmentum. Ann NY Acad Sci (in press)

Beckstead RM (1979) An autoradiographic examination of corticocortical and subcortical projections of the mediodorsal-projection (prefrontal) cortex in the rat. J Comp Neurol 184: 43–62

Benes FM, McSparren J, Bird ED, SanGiovanni JP, Vincent SL (1991) Deficits in small interneurons in prefrontal and cingulate cortex of schizophrenic and schizoaffective patients. Arch Gen Psychiatry 48: 996–1001

Benes FM, Vincent SL, Alsterberg G, Bird ED, SanGiovanni JP (1992) Increased GABA-A receptor binding in superficial layers of schizophrenic cingulate cortex. J Neurosci 12: 924–929

Bennett JP, Leslie CA (1990) NMDA receptors modulate striatal dopamine release and metabolism in vivo: a microdialysis study. Soc Neurosci Abstr 16: 679

Berger B, Verney C (1984) Development of the catecholamine innervation in rat neocortex: morphological features. In: Descarries L, Reader TR, Jasper HH (eds) Monoamine innervation of the cerebral cortex. Alan R Liss, New York, pp 95–121

Berger B, Gaspar P, Verney C (1991) Dopaminergic innervation of the cerebral cortex: unexpected differences between rodents and primates. Trends Neurosci 14: 21–27

Berman KF, Weinberger D (1990) Thre prefrontal cortex in schizophrenia and other neuropsychiatric disease: in vivo physiological correlates of cognitive deficits. In: Uylings HBM, Van Eden CG, De Bruin JPC, Corner MA, Feenstra MPG (eds) The prefrontal cortex: its structure, function, and pathology. Elsevier Science Publishers, Amsterdam, pp 521–538 (Prog Brain Res 85)

Bertolucci-D'Angio M, Serrano A, Driscoll P, Scatton B (1990) Involvement of mesocorticolimbic dopaminergic systems in emotional states. In: Uylings HBM, Van Eden CG, De Bruin JPC, Corner MA, Feenstra MPG (eds) The prefrontal cortex: its structure, function, and pathology. Elsevier Science Publishers, Amsterdam, pp 405–418 (Prog Brain Res 85)

Black IB, Green SC (1975) Postmortem changes in brain catecholamine enzymes. Arch Neurol 32: 47–49

Bleuler E (1950) Dementia praecox or the group of schizophrenias (Zinkin J, trans). International Universities Press, New York

Bolam JP (1984) Synapses of identified neurons in the neostriatum. In: Evered D, O'Connor M (eds) Functions of the basal ganglia. Pittman Press, London, pp 42–57 (Ciba Foundation Symposium 107)

Bouyer JJ, Park DH, Joh TH, Pickel VM (1984) Chemical and structural analysis of the relation between cortical inputs and tyrosine hydroxylase-containing terminals in rat neostriatum. Brain Res 302: 267–275

Braszko JJ, Bannon MJ, Bunney BS, Roth RH (1981) Intra-striatal kainic acid: acute effects on electrophysiological and biochemical measures of nigrostriatal dopaminergic activity. J Pharmacol Exp Ther 216: 289–293

Brog JM, Deutch AY, Zahm DS (1991) Afferent projections to the nucleus accumbens core and shell in the rat. Soc Neurosci Abstr 17: 454

Brown GW, Birley JLP, Wing JK (1972) Influence of family life on the course of schizophrenic illness: a replication. Br J Psychiatry 14: 241–252

Bunney BS, Aghajanian GK (1976) Dopamine and norepinephrine innervated cells in the rat prefrontal cortex: pharmacological differentiation using micro-iontophoretic techniques. Life Sci 19: 1783–1792

Bunney BS, Chiodo LA (1984) Mesocortical dopamine systems: further electrophysiological and pharmacological characterization. In: Descarries L, Reader TR, Jasper HH (eds) Monoamine innervation of the cerebral cortex. Alan R Liss, New York, pp 263–278

Carpenter WT Jr (1991) Psychopathology and common sense: where we went wrong with negative symptoms. Biol Psychiatry 29: 735–737

Carter CJ (1982) Topographical distribution of possible glutamatergic pathways from the frontal cortex to the striatum and substantia nigra in rats. Neuropharmacology 21: 383–393

Carter CJ, Pycock CJ (1980) Behavioral and biochemical effects of dopamine and noradrenaline depletion within the medial prefrontal cortex of the rat. Brain Res 192: 163–176

Cheramy A, Romo R, Glowinski J (1984) Role of corticostriatal glutamatergic neurons in the presynaptic control of dopamine release. In: Sandler M, Feuerstein C, Scatton B (eds) Neurotransmitter interactions in the basal ganglia. Raven Press, New York, pp 133–141

Cheramy A, Romo R, Godehu G, Baruch P, Glowinski J (1986) In vivo presynaptic control of dopamine release in the cat caudate nucleus. II. Facilitatory or inhibitory influence of L-glutamate. Neuroscience 19: 1081–1090

Chesselet M-F (1984) Presynaptic regulation of neurotransmitter release in the brain: facts and hypotheses. Neuroscience 12: 347–375

Christie NJ, Bridge S, James LB, Beart PM (1985) Excitotoxic lesions suggest an aspartatergic projection from rat medial prefrontal cortex to ventral tegmental area. Brain Res 333: 169–172

Clow DW, Jhamandas K (1989) Characterization of L-glutamate action on the release of endogenous dopamine from the rat caudate-putamen. J Pharmacol Exp Ther 284: 722–728

Consolo S, Landinski H, Bianchi S, Ghezzi D (1977) Apparent lack of a dopaminergic-cholinergic link in the rat nucleus accumbens septi-tuberculum olfactorium. Brain Res 135: 255

Cox DWG, Headley MH, Watkins JC (1977) Actions of L- and D-homocysteate in rat CNS: a correlation between low-affinity uptake and the time courses of exciation by microiontophoretically applied L-glutamate analogues. J Neurochem 29: 579–588

Czernansky JG, Murphy GM, Faustman WO (1991) Limbic/mesolimbic connections and the pathogenesis of schizophrenia. Biol Psychiatry 30: 383–400

Deutch AY (1991) Heterogeneity of the prefrontal cortical dopamine system in responsiveness to stress. Soc Neurosci Abstr 17: 529

Deutch AY, Roth RH (1990) The determinants of stress-induced activation of the prefrontal cortical dopamine system. In: Uylings HBM, Van Eden CG, De Bruin JPC, Corner MA, Feenstra MPG (eds) The prefrontal cortex: its structure, function, and pathology. Elsevier Science Publishers, Amsterdam, pp 367–393 (Prog Brain Res 85)

Deutch AY, Cameron DS (1992) Pharmacological characterization of dopamine systems in the nucleus accumbens core and shell. Neuroscience 46: 49–56

Deutch AY, Zahm DS (1992) The current status of neurotensin-dopamine interactions: issues and speculations. Ann NY Acad Sci (in press)

Deutch AY, Tam S-Y, Roth RH (1985) Footshock and conditioned stress increase 3,4-dihydroxyphenylacetic acid (DOPAC) in the ventral tegmental area but not substantia nigra. Brain Res 333: 143–146

Deutch AY, Goldstein M, Baldino F Jr, Roth RH (1988) The telencephalic projections of the A8 dopamine cell group. Ann NY Acad Sci 537: 27–50

Deutch AY, Clark WA, Roth RH (1990) Prefrontal cortical dopamine depletion enhances the responsiveness of mesolimbic dopamine neurons to stress. Brain Res 521: 311–315

Deutch AY, Moghaddam B, Innis R, Krystal JH, Aghajanian GK, Bunney BS, Charney DS (1991a) Mechanisms of action of atypical antipsychotic drugs: implications for novel therapeutic strategies for schizophrenia. Schizophr Res 4: 121–156

Deutch AY, Lee MC, Gillham MH, Cameron D, Goldstein M, Iadorola MJ (1991b) Stress selectively increases Fos protein in dopamine neurons innervating the prefrontal cortex. Cerebral Cortex 1: 273–292

Do KQ, Herrling PL, Streit P, Turski WA, Cuenod M (1986) In vitro release and electrophysiological effects in situ of homocysteic acid, an endogenous N-methyl-(D)-aspartic agonist in the mammalian striatum. J Neurosci 6: 2226–22341

Donoghue JP, Herkenham M (1986) Neostriatal projections from individual cortical fields conform to histochemically distinct striatal compartments in the rat. Brain Res 365: 397–403

Elsworth JD, Deutch AY, Redmond DE Jr, Sladek JR Jr, Roth RH (1990) MPTP reduces dopamine and norepinephrine concentrations in the supplementary motor area and cingulate cortex of the primate. Neurosci Lett 114: 316–312

Errami M, Nieoullon A (1986) Development of a micromethod to study the Na^+-independent L-[^{3}H]glutamic acid binding to rat striatal membranes. I. Biochemical and pharmacological characterization. Brain Res 366: 169–177

Friedemann MN, Gerhardt GA (1990) The effects of excitatory amino acids on dopamine nerve terminals in the neostriatum of the anesthetized rat: an in vivo electrochemical study. Soc Neurosci Abstr 16: 584

Fuxe K, Fredholm BB, Agnati LF, Corrodi H (1978) Dopamine receptors and ergot drugs. Evidence that an ergolene derivative is a differential agonist at subcortical limbic dopamine receptors. Brain Res 146: 295–311

Gariano RF, Groves PM (1988) Burst firing induced in midbrain dopamine neurons by stimulation of the medial prefrontal and anterior cingulate cortices. Brain Res 462: 194–198

Gerfen CR (1989) The neostriatal mosaic: striatal patch-matrix organization is related to cortical lamination. Science 246: 385–388

Gerfen CR, Herkenham M, Thibault J (1987) The neostriatal mosaic. II. Patch- and matrix-directed mesostriatal dopaminergic and non-dopaminergic systems. J Neurosci 7: 3915–3934

Giorguieff MF, Kemel ML, Glowinski J (1977) Presynaptic effect of L-glutamatic acid on the release of dopamine in rat striatal slices. Neurosci Lett 6: 73–77

Girault JA, Spampinato U, Glowinski J, Besson MJ (1986) In vivo release of [^{3}H]γ-aminobutyric acid in the rat neostriatum. II. Opposing effects of D1 and D2 dopamine receptor stimulation in the dorsal caudate putamen. Neuroscience 19: 1109–1117

Girault JA, Spampinato U, Savaki HE, Glowinski J, Besson MJ (1986) In vivo release of [^{3}H]γ-aminobutyric acid in the rat neostriatum. I. Characterization and topographical heterogeneity of the effects of dopaminergic and cholinergic agents. Neuroscience 19: 1100–1108

Goldstein M, Deutch AY (1992) Dopaminergic mechanisms in the pathogenesis of schizophrenia. FASEB J 6:2413–2421

Goldstein M, Lieberman AN, Helmer E, Koslow M, Ransohoff J, Elsworth JD, Roth RH, Deutch AY (1988) Biochemical analysis of caudate nucleus biopsy samples from parkinsonian patients. Ann Neurol 24: 685–688

Goldman-Rakic PS, Leranth C, Williams SM, Mons N, Geffard M (1989) Dopamine synaptic complex with pyramidal neurons in primate cerebral cortex. Proc Natl Acad Sci USA 86: 9015–9019

Grace AA (1991) Phasic versus tonic dopamine release and the modulation of dopamine system responsivity: a hypothesis for the etiology of schizophrenia. Neuroscience 41: 1–24

Graybiel AM (1984) Correspondence between the dopamine islands and striosomes of the mammalian striatum. Neuroscience 13: 1157–1187

Gutman CR, Booze RM, Davis JN (1987) Benefits of rapid vs. delayed autopsy in human brain catecholamine axonal morphology. Soc Neurosci Abstr 13: 437

Haroutunian V, Knott P, Davis KL (1988) Effects of mesocortical dopaminergic lesions upon subcortical dopaminergic function. Psychopharmacol Bull 24: 341–344

Hefti F, Melamed E, Wurtman RJ (1980) Partial lesions of the nigrostriatal system in rat brain: biochemical characterization. Brain Res 195: 123–137

Heimer L, Zahm DS, Churchill L, Kalivas PW, Wohltmann C (1991) Specificity in the projection patterns of the accumbal core and shell. Neuroscience 41: 89–126

Herkenham M (1991) Mismatches between neurotransmitter and receptor localizations: implications for endocrine functions in brain. In: Fuxe K, Agnati LF (eds) Volume transmission in the brain: novel mechanisms for neural transmission. Raven Press, New York, pp 63–87

Imperato A, Honore T, Jensen LH (1990) Dopamine release in the nucleus caudatus and in nucleus accumbens in under glutamatergic control through non-NMDA receptors: a study in freely moving rats. Brain Res 530: 223–228

Jaskiw GE, Weinberger DR, Crawley JN (1991) Microinjection of apomorphine into the prefrontal cortex of the rat reduces dopamine metabolite concentrations in microdialysate from the caudate nucleus. Biol Psychiatry 29: 703–706

Jhamandas K, Marien M (1987) Glutamate-evoked release of endogenous brain dopamine: inhibition by an excitatory amino acid antagonist and an enkephalin analogue. Br J Pharmacol 90: 641–650

Joyce EM, Stinus L, Iversen SD (1983) Effect of injections of 6-hydroxydopamine into either nucleus accumbens septi or frontal cortex on spontaneous and drug-induced activity. Neuropharmacology 9: 1141–1145

Julou L, Scatton B, Glowinski J (1976) Acute and chronic treatments with neuroleptics: similarities and differences in their action on nigrostriatal, mesolimbic, and mesocortical dopaminergic neurons. Adv Biochem Psychopharmacol 16: 617–624
Kalivas PW, Duffie P (1990) Effect of acute and daily neurotensin and enkephalin treatment on extracellular dopamine in the nucleus accumbens. J Neurosci 10: 2940–2949
Keefe KA, Zigmond MJ, Abercrombie ED (1990) In vivo evidence that basal dopamine efflux in striatum is not regulated by endogenous excitatory amino acids. Soc Neurosci Abstr 16: 522
Kilpatrick IC, Mozely LS (1986) An intial analysis of the regional distribution of excitatory sulphur-containing amino acids in the rat brain. Neurosci Lett 72: 189–193
Kim JM, Hassler R, Haus P, Paik KS (1977) Effect of frontal cortical ablation on striatal glutamic acid level in the rat. Brain Res 132: 370–374
Knopfel T, Zeise ML, Cuenod M, Ziegelgansberger W (1987) L-Homocysteic acid but not L-glutamate is an endogenous N-methyl-D-aspartic acid receptor preferring agonist in rat neocortical neurons in vitro. Neurosci Lett 81: 188–192
Knorr AM, Deutch AY, Roth RH (1989) The anxiogenic β-carboline FG 7142 increases in vivo and in vitro tyrosine hydroxylation in the prefrontal cortex. Brain Res 495: 355–361
Krebs MO, Desce JM, Kemel ML, Gauchy C, Godeheu G, Cheramy A, Glowinski J (1991) Glutamatergic control of dopamine release in the rat striatum: evidence for presynaptic N-methyl-D-aspartate receptors on dopaminergic nerve terminals. J Neurochem 56: 81–85
Leccese AP, Lyness WH (1987) Lesions of dopamine neurons in the medial prefrontal cortex: effects on self-administration of amphetamine and dopamine synthesis in the brain of the rat. Neuropharmacology 26: 1303–1308
Lehmann J, Tsai C, Wood PL (1988) Homocysteic acid as a putative excitatory amino acid neurotransmitter. I. Postsynaptic characteristics at N-methyl-D-aspartate-type receptors on striatal cholinergic interneurons. J Neurochem 51: 1765–1770
Lonart G, Zigmond MJ (1991) High glutamate concentrations evoke Ca^{++}-independent dopamine release from striatal slices: a possible role of reverse dopamine transport. J Pharmacol Exp Ther 256: 1132–1138
Louilot A, Taghzouti K, Demineire JM, Simon H, Le Moal M (1987) Dopamine and behavior: functional and theoretical considerations. In: Sandler M, Feuerstein C, Scatton B (eds) Neurotransmitter interactions in the basal ganglia. Raven Press, New York
Louilot A, Le Moal M, Simon H (1989) Opposite influences of dopaminergic pathways to the prefrontal cortex or the septum on dopaminergic transmission in the nucleus accumbens. An in vivo voltammetric study. Neuroscience 29: 45–56
Mantz J, Thierry AM, Glowinski J (1989) Effect of noxious tail pinch on the discharge rate of mesocortical and mesolimbic dopamine neurons: selective activation of the mesocortical system. Brain Res 476: 377–381
Marien M, Brien J, Jhamandas K (1983) Regional release of [^{3}H]dopamine from rat brain in vitro: effects of opiates on release induced by potassium, nicotine, and L-glutamic acid. Can J Physiol Pharmacol 61: 43–60
Martin-Iversen MT, Szostak C, Fibiger HC (1986) 6-Hydroxydopamine lesions of the medial prefrontal cortex fail to influence intravenous self-administration of cocaine. Psychopharmacology 88: 310–314
Mayer ML, Viklicky L (1989) Concanavalin A selectively reduces desensitization of mammalian neuronal quisqualate receptors. Proc Natl Acad Sci USA 86: 1411–1415
McGeer PL, McGeer EG, Scherrer U, Singh K (1977) A glutamatergic cortico-striatal path? Brain Res 128: 369–373
McGeorge AJ, Faull RLM (1989) The organization of the projection from the cerebral cortex to the striatum in the rat. Neuroscience 29: 503–537

Moghaddam B, Bunney BS (1990a) Acute effect of typical and atypical antipsychotic drugs on the release of dopamine from the prefrontal cortex, nucleus accumbens, and striatum of the rat: an in vivo microdialysis study. J Neurochem 54: 1755–1760
Moghaddam B, Gruen RJ (1991) Do endogenous excitatory amino acids influence striatal dopamine release? Brain Res 544: 329–330
Moghaddam B, Gruen RJ, Roth RH, Bunney BS, Adams RN (1990b) Effect of L-glutamate on the release of striatal dopamine: in vivo dialysis and electrochemical studies. Brain Res 518: 55–60
Morgan JI, Curran T (1986) The role of ion flux in the control of c-fos. Nature 322: 552–555
Nicholson IR, Neufeld RWJ (1989) Forms and mechanisms of susceptibility to stress in schizophrenia. In: Neufeld RWJ (ed) Advances in the investigation of psychological/stress. Wiley, New York, pp 392–420
Oades RD, Tagzhouti K, Rivet J-M, Simon H, Le Moal M (1986) Locomotor activity in relation to dopamine and noradrenaline in the nucleus accumbens, septal, and frontal areas: a 6-hydroxydopamine study. Neuropsychobiology 16: 37–42
Olson L, Seiger A, Fuxe F (1972) Heterogeneity of striatal and limbic dopamine innervation: highly fluorescent islands in developing and adult rats. Brain Res 44: 283–288
Palmer AM, Hutson PH, Lowe SL, Bowen DM (1989) Extracellular concentration of aspartate and glutamate in rat neostriatum following chemical stimulation of frontal cortex. Exp Brain Res 75: 659–663
Passingham RE, Myers C, Rawlins N, Lightfoot V, Fearn S (1988) Premotor cortex in the rat. Behav Neurosci 102: 101–109
Payson MM, Donzanti BA (1990) Effects of excitatory amino acids on in vivo dopamine release and metabolism in the nucleus accumbens. Soc Neurosci Abstr 16: 584
Penit-Soria J, Audinat E, Crepel F (1987) Excitation of rat prefrontal cortical neurons by dopamine: an in vitro eletrophysiological study. Brain Res 425: 363–374
Peterson OT, St Mary JS, Harding NR (1987) Cis-flupenthixol antagonism of the rat prefrontal cortex neuronal response to apomorphine and ventral tegmental area input. Brain Res Bull 18: 723–729
Pycock CJ, Carter CJ, Kerwin RW (1980a) Effect of 6-hydroxydopamine lesions of the medial prefrontal cortex on neurotransmitter systems in subcortical sites in the rat. J Neurochem 34: 91–99
Pycock CJ, Kerwin RW, Carter CJ (1980b) Effect of lesion of cortical dopamine terminals on subcortical dopamine receptors in rats. Nature 286: 74–77
Retaux S, Besson MJ, Penit-Soria J (1991) Synergism between D_1 and D_2 dopamine receptors in the inhibition of the evoked release of [^{3}H]GABA in the rat prefrontal cortex. Neuroscience 43: 323–329
Roberts PJ, Anderson SD (1979) Stimulatory effect of L-glutamate and related amino acids on [^{3}H]dopamine release from rat striatum: an in vitro model for glutamate actions. J Neurochem 32: 1539–1541
Rosin DJ, Clark WA, Goldstein M, Roth RH, Deutch AY (1992) Effects of 6-hydroxydopamine lesions of the prefrontal cortex on tyrosine hydroxylase activity in subcortical dopamine systems of the rat. Neuroscience (in press)
Sagar SM, Sharp FR, Curran T (1988) Expression of c-fos protein in brain: metabolic mapping at the cellular level. Science 240: 1328–1331
Samuel D, Errami M, Nielullon A (1990) Localization of N-methyl-D-aspartate receptors in the rat striatum: effects of specific lesions on the [^{3}]3-(2-carboxypiperazin-4-yl)propyl-1-phosphonic acid binding. J Neurochem 54: 1926–1933
Seguela P, Watkins KC, Descarries L (1988) Ultrastructural features of dopamine axon terminals in the anteromedial and the suprarhinal cortex of adult rat. Brain Res 442: 11–22
Sesack SR, Bunney BS (1989) Pharmacological characterization of the receptor

mediating electrophysiological responses to dopamine in the rat medial prefrontal cortex: a microiontophoretic study. J Pharmacol Exp Ther 248: 1323–1333
Sesack SR, Deutch AY, Roth RH, Bunney BS (1989) The topographical organization of the efferent projections of the medial prefrontal cortex in the rat: an anterograde tract-tracing study using Phaseolus vulgaris leucoagglutinin. J Comp Neurol 290: 213–242
Sheng M, Greeberg ME (1990) The regulation and function of c-fos and other immediate early genes in the nervous system. Neuron 4: 477–485
Shepard PD, German DC (1984) A subpopulation of mesocortical dopamine neurons possess autoreceptors. Eur J Pharmacol 98: 455–456
Spencer HJ (1976) Antagonism of cortical excitation of striatal neurons by glutamic acid diethylester: evidence for glutamic acid as an excitatory transmitter in the rat striatum. Brain Res 102: 91–101
Swanson LW (1982) The projections of the ventral tegmental area and adjacent areas: a combined retrograde tracer and immunofluorescence study of the rat. Brain Res Bull 9: 321–353
Tandon R, Greden JF (1991) Negative symptoms of schizophrenia: the need for conceptual clarity. Biol Psychiatry 30: 321–325
Tassin JP, Studler JM, Herve D, Blanc G, Glowinski J (1986) Contribution of noradrenergic neurons to the regulation of dopaminergic (D_1) receptor denervation supersensitivity in rat prefrontal cortex. J Neurosci 46: 243–248
Thierry AM, Tassin JP, Blanc G, Glowinski J (1976) Selective activation of the mesocortical dopamine system by stress. Nature 263: 242–244
Thierry AM, Le Douarin C, Penit J, Ferron A, Glowinski J (1986) Variation in the ability of neuroleptics to block the inhibitory influence of dopaminergic neurons on the activity of cells in the rat prefrontal cortex. Brain Res Bull 16: 155–160
Thierry A-M, Godbut R, Mantz J, Glowinski J (1990) Influence of the ascending monoaminergic systems on the activity of the rat prefrontal cortex. In: Uylings HBM, Van Eden CG, De Bruin JPC, Corner MA, Feenstra MPG (eds) The prefrontal cortex: its structure, function, and pathology. Elsevier Science Publishers, Amsterdam, pp 377–366 (Prog Brain Res 85)
Vaughn C, Leff JP (1976) The influence of family and social factors on the course of psychiatric illness. Br J Psychiatry 129: 129–137
Weinberger DR (1987) Implications of normal brain development for the pathogenesis of schizophrenia. Arch Gen Psychiatry 44: 660–669
Westerink BHC, de Vries JB (1989) On the mechanism of neuroleptic induced increase in striatal dopamine release: brain dialysis provides direct evidence for mediation by autoreceptors localized on nerve terminals. Neurosci Lett 99: 197–202
Wong M-L, Deutch AY (1991) Localization of Glu-A receptor mRNA in the substantia nigra and ventral tegmental area. Soc Neurosci Abstr 17: 793
Young AMJ, Bradford HF (1986) Excitatory amino acid neurotransmitters in the corticostriate pathway: studies using intracerebral microdialysis in vivo. J Neurochem 47: 1399–1404
Zahm DS (1991) Compartments in rat dorsal and ventral striatum revealed following injection of 6-hydroxydopamine into the ventral mesencephalon. Brain Res 552: 164–169 1991.
Zahm DS, Heimer L (1990) Two transpallidal pathways originating in rat nucleus accumbens. J Comp Neurol 302: 437–446
Zigmond MJ, Abercrombie ED, Berger TW, Grace AA, Stricker EM (1990) Compensations after lesions of central dopaminergic neurons: some basic and clinical implications. Trends Neurosci 13: 290–295

Author's address: Dr. A. Y. Deutch, Department of Psychiatry, Yale University School of Medicine, Connecticut Mental Health Center, 34 Park Street, New Haven, CT 06508, U.S.A.

J Neural Transm (1992) [Suppl] 36: 91–131

The depolarization block hypothesis of neuroleptic action: implications for the etiology and treatment of schizophrenia

A. A. Grace

Departments of Behavioral Neuroscience and Psychiatry, University of Pittsburgh, Pittsburgh, PA, U.S.A.

Summary. Antipsychotic drugs are known to block dopamine receptors soon after their administration, resulting in an increase in dopamine neuron firing and dopamine turnover. Nonetheless, antipsychotic drugs must be administered repeatedly to schizophrenics before therapeutic benefits are produced. Recordings from dopamine neurons in rats have revealed that chronic antipsychotic drug treatment results in the time-dependent inactivation of dopamine neuron firing via over-excitation, or depolarization block. Furthermore, the clinical profile of the response to antipsychotic drugs appears to correspond to the dopamine system affected: antipsychotic drugs that exert therapeutic actions in schizophrenics inactivate dopamine neuron firing in the limbic-related ventral tegmental area, whereas drugs that precipitate extrapyramidal side effects cause depolarization block of the motor-related substantia nigra dopamine cells.

One factor that remains unresolved with regard to the actions of antipsychotic drugs is the relationship between dopamine turnover and depolarization block — i.e., why does a significant level of dopamine release or turnover remain after antipsychotic drug treatment if dopamine cells are no longer firing? We addressed this question using an acute model of neuroleptic-induced depolarization block. In this model, dopamine cells recorded in rats one month after partial dopamine lesions could be driven into depolarization block by the acute administration of moderate doses of haloperidol. However, similar doses of haloperidol, which were effective at increasing dopamine levels in the striatum of intact rats, failed to change dopamine levels in lesioned rats. This is consistent with a model in which neuroleptic drugs exert their therapeutic effects in schizophrenics by causing depolarization block in DA cells, thereby preventing further activation of dopamine neuron firing in response to external stimuli. Thus, attenuating the responsivity of the dopamine system to stimuli may be more relevant to the therapeutic actions of antipsychotic drugs than receptor blockade or decreases in absolute levels of dopamine, which could presumably be circumvented by homeostatic adaptations in this highly plastic system.

Introduction

Schizophrenia is reported to affect 1% of the population worldwide (Torrey, 1987). Although it is not the most prevalent of the major psychiatric disorders, the early age of onset and chronic debilitating nature of schizophrenia cause it to be among the most devastating illnesses in its effects on afflicted individuals and their families. However, a major inroad into reducing the impact of schizophrenia was realized with the relatively recent discovery of antipsychotic drugs and their introduction into clinical psychiatry, which has served to transform the long-term treatment and care of schizophrenic patients. In the decades before the use of antipsychotic drugs, the population of patients in U.S. state and county mental hospitals was growing at an alarming rate, with some reports showing an increase in patients from less than 100,000 in 1900 to more than 550,000 by 1955 (Davis, 1985). Since no effective treatment was available, the patient with schizophrenia had little hope for release. The hospitals were capable of providing these patients with only a custodial level of care, with the result that two-thirds of the schizophrenics spent most of their lives confined to insane asylums. This rather dismal picture was dramatically transformed with the implementation of antipsychotic drug therapy. In a short period of time, hospitals were able to change their approach to the schizophrenic patient: instead of serving merely as a repository for the permanent confinement of schizophrenics, they were now able to employ a positive approach to treatment. This trend progressed over the years with the consequence that at present almost 95% of those treated with antipsychotic drugs can be discharged. This development contributed to the approximately 80% decline in patient population since 1955 (Stevens, 1991).

The efficacy of antipsychotic drugs in the treatment of schizophrenia is well-accepted, having been confirmed in numerous clinical trials (e.g., Klein and Davis, 1969). In an NIMH collaborative study it was shown that, of the schizophrenics treated with antipsychotic drugs, almost 70% showed significant improvement whereas only 10% failed to show any response; none of the subjects exhibited a worsening of symptoms. In contrast, of the schizophrenics treated with placebo, only 25% showed any spontaneous improvement during the period of treatment, whereas the majority exhibited no change or increased debilitation. Furthermore, those stricken with schizophrenia could expect a significantly lower rate of relapse if maintained on antipsychotic drugs. Treatment with antipsychotic drugs may also play a role in the long-term prognosis of schizophrenia, in that recent epidemiological studies have provided convincing evidence that antipsychotic drugs may actually contribute to a slowing in the progress of the disease (Wyatt, 1991; May et al., 1976).

As with many other drugs used to treat psychiatric disorders, the discovery of the first antipsychotic drug occurred by accident rather than by design. Historical accounts (Snyder, 1980) reveal that these drugs originated during an effort to develop a less sedating derivative of the phenothiazine promethazine, which at that time was widely used as an antihistaminergic agent (Charpentier, 1950). The resultant drug, which became known as chlorpromazine, was found to be too sedating to be of much therapeutic value as an antihistamine. As a consequence of its sedative action, this drug was initially used by the French anesthesiologist-surgeon Laborit (Laborit et al., 1952) to decrease autonomic stress during surgical procedures. The observation that it produced a type of tranquilization led to trials in psychiatric patients as a potential tool to suppress violent episodes or agitation. This introduction of chlorpromazine into the psychiatric clinic enabled Delay and colleagues (1952) to discover that, in a subgroup of patients, chlorpromazine induced more than mere sedation: it actually appeared to reduce the core symptoms of schizophrenia.

Despite the rapid implementation and dramatic benefits derived from the use of chlorpromazine in the clinic, the nature of its effects on brain systems remained a mystery for almost ten years. The discovery of brain dopamine (DA) systems in 1958 (Carlsson et al., 1958), however, supplied the missing information necessary to analyze the actions of this compound. In a now classic study by Carlsson and Lindqvist (1963), injection of rats with antipsychotic drugs was found to produce an increase in DA metabolite levels in the brain. In a rather profound intuitive leap, Carlsson proposed that this effect was due to a blockade of DA receptors, with the resultant increase in DA metabolites occurring secondary to a feedback increase in DA neuron firing. Although it required another ten years before this idea could be adequately evaluated, its main components have been substantiated experimentally — antipsychotic drugs were found to: displace DA from its receptor (Seeman et al., 1975) with an affinity that correlates with their potency in the clinic (Creese et al., 1976; Seeman et al., 1976), accelerate the firing rate of identified DA neurons, and block the inhibitory actions of the psychotomimetic drug amphetamine on DA cell firing (Bunney et al., 1973).

The dopamine hypothesis of schizophrenia: shortcomings of the model

Of the various neurotransmitter systems in the brain that have been proposed to play a role in the etiology of schizophrenia, only the DA system has maintained the long-term interest of investigators in this field. As related above, the most compelling evidence for the involvement of DA in this disorder was derived from studies related to the central actions of antipsychotic drugs. Thus, despite the variety of neurotransmitter systems that can be influenced by different classes of antipsychotic drugs, all therapeutically effective drugs to date have had the common property of blocking DA receptors, particularly those of the D_2 subtype (Kebabian and Calne, 1979; Peroutka and Snyder, 1980). The ability of drugs that amplify the actions of the DA system, such as L-DOPA or D-amphetamine, to exacerbate schizophrenia in human patients and to cause schizophreniform behavior in normal subjects has strengthened this association (Angrist et al., 1971; Connell, 1958). Thus, the combined impact of the discovery of the DA blocking properties of antipsychotic drugs and the DA releasing properties of D-amphetamine (Carlsson et al., 1966; Glowinski et al., 1966) served to ensconce the DA hypothesis of schizophrenia into biological psychiatry.

The DA hypothesis, in its simplest form, is based on the idea that schizophrenia results from a hyperactivity of the DA system (Snyder, 1972; Van Rossum, 1966). However, when one attempts to determine the basic physiological deficit that underlies schizophrenia, little data can be found to substantiate a dysfunctional DA system as the primary causative factor in this disorder. This is particularly significant when considered in light of what is known about the resilience of the DA system in disorders in which the role of DA is clearly established, such as in Parkinson's disease. Thus, evidence from studies in human Parkinsonian patients (Bernheimer et al., 1973; Lloyd et al., 1975) as well as in rats (Ungerstedt, 1971; Zigmond and Stricker, 1973) has revealed that this neurotransmitter system is capable of a remarkable degree of plasticity, in that depletions of 80% or more of striatal DA can be compensated for sufficiently by DA system up-regulation to allow behavioral recovery of the animal (Zigmond et al., 1989). Conversely, continuous stimulation of DA receptors by DA agonists will cause a down-regulation of the DA system (Friedhoff et al., 1977; Scatton and Worms, 1978). Under these conditions, one may expect that rather extensive pathological changes would have to occur in the DA system in order to circumvent this powerful homeostatic restorative influence before a dysfunctional state, particularly one as profound as schizophrenia, could be produced. Indeed, even the nature of the therapeutic effect exerted by antipsychotic drugs is problematic: although antipsychotic drugs can achieve a high level of receptor occupancy in humans within hours after acute oral administration (Sedvall et al., 1986), clinical investigations have consistently shown that antipsychotic drugs must be administered to schizophrenics for weeks before their therapeutic benefits can be fully realized (e.g., Johnstone et al., 1978). Furthermore, this time-dependent increase in therapeutic efficacy is paralleled by the development of DA receptor supersensitivity (Burt et al., 1977), which would be expected to offset the DA blocking actions of antipsychotic drugs. Thus, it is unclear what long-term therapeutic value could be derived from blockade of DA receptors when one takes into account: 1) the absence of evidence for a substantial pathological increase in the activity of the DA system in schizophrenics, 2) the minimal effect of acute DA receptor blockade on schizophrenic symptoms, and 3) the formidable ability of the DA system to compensate for chronic changes in DA receptor stimulation.

Differences in the effects of acute and chronic neuroleptic treatment

The acute actions of antipsychotic drugs have been well-described in animals and, for the most part, parallel their effects in humans. Many of the acute effects on motor behavior produced by antipsychotic drugs like haloperidol can be observed within minutes following their administration, and consist of a dose-dependent decrease in spontaneous movement culminating in akinesia and catalepsy (Janssen et al., 1965; Munkvad et al., 1968). Furthermore, haloperidol rapidly reverses the stereotyped behavior elicited by DA agonists, such as apomorphine or amphetamine (del Rio and Fuentes, 1969; Randrup and Munkvad, 1965). The effects of antipsychotic drugs on biochemical measures of DA function also are produced soon after their administration. Thus, within minutes after an injection of haloperidol, an accumulation of DA in the extracellular fluid of the striatum is observed (Imperato and Dichiara, 1985; Zetterstrom et al., 1985). This is believed to be analogous to the increase in CSF or plasma levels of the DA metabolite homovanillic acid (HVA) in schizophrenics in response to the acute administration of neuroleptics (Bowers, 1973). This is in marked contrast to the effects produced during chronic administration of antipsychotic drugs to patients, in which the DA metabolites show a progressive decline from the elevated levels induced acutely to levels approaching those of control, unmedicated patients (Post and Goodwin, 1975). In fact, clinical studies have shown that the decline in HVA occurred in parallel with the development of the antipsychotic effects of the treatment (Pickar et al., 1984, 1986; Davila et al., 1988a). Thus, the initial increase in HVA produced by the acute administration of antipsychotic drugs was replaced by the development of tolerance to neuroleptic-induced increases in DA turnover, which appeared to correlate with the emergence of antipsychotic effects.

The role of dopamine neuron depolarization block in neuroleptic action

Studies of DA metabolism during the course of antipsychotic drug administration suggested that the time-dependent increase in the therapeutic effects of these drugs was contingent on the development of a blunted response within the DA system to the medication. However, it was unclear what changes could be induced in the system to produce this blunted response to neuroleptics. Since the increase in HVA produced by acute neuroleptic administration was thought to reflect a feedback excitation of DA neuron firing (Carlsson and Lindqvist, 1963; Bunney et al., 1973), one may predict that DA cell firing should show tolerance to chronic neuroleptic administration. In order to address this issue, we developed a method that was capable of assessing the changes in activity across a population of DA neurons during the course of chronic drug administration (Bunney and Grace, 1978). This was done by taking advantage of the anatomical organization of DA neurons in the substantia nigra (SN): a compact but thin sheet of cells approximately 5–8 neurons thick extending across a comparatively large mediolateral and rostroventral expanse. By passing an extracellular recording electrode through this region twelve times in a predefined pattern and recording the firing of each DA neuron encountered, we were able to obtain a reproducible measure of the relative number of DA cells firing per electrode track, their average firing rate, and their firing pattern.

We found that acute administration of the antipsychotic drug haloperidol activated DA neuron firing as reflected by: 1) an increase in the average firing rate of DA neurons, 2) a switch in DA cell firing from a single-spiking mode to a burst-firing pattern (Bunney et al., 1973; Grace and Bunney, 1984a,b), which increases the amount of DA released per spike (Gonon, 1988), and 3) an increase in the average number of spontaneously firing DA neurons recorded per electrode track due to the activation of normally quiescent DA neurons (Bunney and Grace, 1978). However, repeated treatment of rats with haloperidol for three weeks or more resulted in a *decrease* in the number of spontaneously active DA neurons observed. In fact, electrophysiological studies of the activity across the population of DA neurons revealed that the majority of neurons were now inactive and did not fire action potentials. Furthermore, this quiescent state appeared to be qualitatively different from that present in non-firing DA neurons in control rats, since cell activity could not be reinstated by application of neurotransmitter substances that normally activate DA cell firing (Bunney and Grace, 1978). Thus, this finding provided a potential explanation for the decrease in DA turnover produced during chronic neuroleptic treatment, although the cause of this electrophysiological change was unclear.

Since acute administration of neuroleptics excited DA neurons, the possibility emerged that after chronic treatment these neurons were not in a hyperpolarized, inactive state, but were instead depolarized to such an extent that their spike-generating mechanism was inactive. A comparable condition, known as depolarization block, had been described in the spinal cord in response to the direct application of large quantities of excitatory amino acids to individual neurons (Curtis et al., 1960), but had not been observed in any system under normal conditions or in response to therapeutically relevant doses of drugs. Nonetheless, systemic administration of the DA agonist apomorphine as well as direct, iontophoretic application of the inhibitory neurotransmitter GABA, both of which inhibit DA neurons in control rats (Aghajanian and Bunney, 1974; Grace and Bunney, 1979, 1985b; Skirboll et al., 1979), were found to activate these "silent" DA neurons in chronic haloperidol-treated rats (Bunney and Grace, 1978). This effect was not due to a change in the manner by which these compounds affected DA neurons, since application of additional quantities of these agonists decreased the firing rate of these recently activated DA neurons.

The neuroleptic-induced inhibition of spontaneous cell firing was initially termed "depolarization inactivation" (Bunney and Grace, 1978). The implementation of *in vivo* intracellular recordings from identified DA neurons (Grace and Bunney, 1980a,1983a) enabled us to characterize DA cell electrophysiological properties, such as the membrane potential, membrane conductance, and spike generating mechanisms (Grace and Bunney, 1983a–c, 1984a,b), in order to determine the primary factors regulating DA cells and how these may influence the development of depolarization block. Using this technique, we confirmed that DA neurons in chronic haloperidol-treated rats were tonically depolarized to the point where their spike generating region was inactivated, and could be restored to a spontaneously firing state if the membrane of the DA cell was hyperpolarized by current injection or by systemic administration of DA agonists (Grace and Bunney, 1986). Since this inactivation of spike generation in DA neurons secondary to tonic membrane depolarization was found to closely approximate the depolarization-induced blockade of spiking in spinal motoneurons cited above, it is now more appropriately referred to as "depolarization block" (Grace and Bunney, 1986).

The experiments detailing the time-dependent electrophysiological effects of antipsychotic drugs on DA neurons provided a tentative explanation for the onset of tolerance, but it was unclear whether this electrophysiological change was directly related to the therapeutic actions of antipsychotic drugs. Evidence that had been accumulating at this time suggested that midbrain DA systems could be functionally subdivided, with the nigrostriatal system predominantly involved in motor function and the extrapyramidal effects of neuroleptics (Marsden et al., 1975), and the DA projections innervating central limbic and cortical regions possibly playing a greater role in the cognitive disturbances of schizophrenics (Snyder, 1973; Matthysse, 1973; Stevens, 1973). Accordingly, White and Wang (1983a) compared the effects produced by repeated haloperidol administration on DA cell firing in SN DA neurons projecting to the striatum with its effects on the ventral tegmental area (VTA) DA neurons innervating limbic and cortical regions, and found that not only did haloperidol cause VTA DA neurons to enter depolarization block, it did so to a greater extent than was observed in the SN. Furthermore, depolarization block required three weeks to develop in both the extrapyramidal DA system and the mesolimbic DA system.

Classical versus atypical antipsychotic drugs: insights into the origin of the extrapyramidal side-effects of neuroleptics

Depolarization block of DA neuron firing provided a potential mechanism to explain the development of tolerance in the DA system during neuroleptic administration. The finding that classical neuroleptics, which are known to induce extrapyramidal side effects in addition to possessing antipsychotic efficacy, caused depolarization block in both the extrapyramidal and limbic DA system provided additional support for this proposed mechanism of action of neuroleptic drugs. This combined action on both the limbic- and motor-related DA systems is a common characteristic of classical antipsychotic drugs; indeed, the term "antipsychotic drug" was originally used interchangeably with the term "neuroleptic," since it was believed that the extrapyramidal effects of these drugs were a characteristic feature present in all therapeutically effective antipsychotic drugs (Delay et al., 1955; Flugel, 1956; Deniker, 1960). This identity of action was discounted with the introduction of the so-called "atypical" antipsychotic agents, which were reported to be devoid of extrapyramidal side-effects (Stille and Hippius, 1971). Repeated treatment of rats with clozapine or other atypical antipsychotic drugs induced depolarization block in the mesolimbic DA neurons but not in the nigrostriatal system (Chiodo and Bunney, 1983; White and Wang, 1983b). In contrast, chronic administration of the DA antagonist metoclopramide, which causes extrapyramidal effects but, at low doses, is not an effective antipsychotic drug (Wood, 1977; Lavy et al., 1978), produced depolarization block only in the SN DA system (White and Wang, 1983b).

These studies support a role for depolarization block of DA cell activity in the SN in the extrapyramidal side-effects of antipsychotic drugs, and depolarization block in the VTA in their therapeutic actions. Furthermore, since the DA neurons projecting to the frontal cortex are not driven into depolarization block by chronic neuroleptics (Chiodo and Bunney, 1983), the therapeutic effects of these drugs appear to be due to their effects on the mesolimbic DA neurons within the VTA. Indeed, there are several additional reasons to doubt that the mesocortical DA system is involved in the therapeutic actions of antipsychotic drugs: 1) the hypofrontality present in schizophrenics (Ingvar and Franzen, 1974; Weinberger et al., 1986) is not reversed during the course of treatment with therapeutically effective antipsychotic drugs (Buchsbaum et al., 1987; Pettegrew et al., 1989,1991), and 2) although the efficacy of antipsychotic drugs is correlated with their affinity for D_2 receptors (Seeman et al., 1987; Farde et al., 1989), there is little D_2-specific binding in the frontal cortex of primates (Lidow et al., 1989) or humans (De Keyser et al., 1988). Thus, antipsychotic drugs seem more likely to exert their therapeutic effects subcortically on the mesolimbic DA system.

The selective actions of clozapine on mesolimbic DA cells is consistent with the absence of extrapyramidal side effects associated with this drug. However, it is not clear what properties of clozapine limit its actions to the VTA DA neuronal system. One proposal is that this relative selectivity is associated with the ability of clozapine to preferentially block DA actions on neurons in the limbic system (Bunney and Aghajanian, 1978), which is consistent with its higher affinity for the D_3 (Sokoloff et al., 1990) or D_4 (Van Tol et al., 1991) subtypes of DA receptors that are preferentially located in the limbic regions of the DA projection system. Other studies suggest that this selectivity of clozapine results instead from its "non-specific" actions on muscarinic cholinergic receptors and/or α_1-adrenergic receptors (Leysen, 1982).

The systems involved in the differential effects of classic versus atypical antipsychotic drugs were analyzed in electrophysiological experiments in which rats were treated repeatedly with a combination of haloperidol and either the α_1-adrenergic blocker prazosin or the anticholinergic drug trihexylphenidyl (Chiodo and Bunney, 1985). In each case, the drug combination produced a physiological profile similar to that induced by clozapine: selective depolarization block of the VTA DA neurons. These results are consistent with clinical findings showing that co-administration of haloperidol and anticholinergic drugs reduces the incidence and the severity of extrapyramidal side-effects (Fann and Lake, 1976; Manos et al., 1981). Although the interaction of α-adrenergic drugs with neuroleptics have yet to be evaluated in clinical trials, suggestive evidence shows that clonidine may be effective in reducing chronic neuroleptic-induced tardive dyskinesia in monkeys (Nishikawa et al., 1983b).

Systems involved in the induction of depolarization block

The decrease in biochemical response to haloperidol during repeated administration (i.e., tolerance) is consistent with what one might predict based on simple pharmacological models of drug action; i.e., repeated treatment with a neurotransmitter agonist or antagonist should induce compensatory changes that serve to counteract the effects of the drug. However, it is apparent that repeated treatment with haloperidol actually exerts a progressively *greater* effect with respect to the depolarization of DA neurons, eventually culminating in depolarization block. There are at least two sites where haloperidol may potentially exert a major influence on DA cell activity: 1) on DA cell body autoreceptors in the SN and VTA (Aghajanian and Bunney, 1977), and 2) on DA receptors in the striatum, where blockade can activate striatonigral feedback loops (Grace and Bunney, 1985a).

Evidence reported to date suggests that striatonigral feedback pathways are more directly involved in the development of depolarization block in the SN. Thus, kainic acid-induced unilateral lesions of the striatum prior to chronic haloperidol administration prevented the development of DA cell depolarization block only on the lesioned side (Bunney and Grace, 1978, 1979). Furthermore, although an acute transection of the brain between the striatum and the substantia nigra has little influence on the basal activity of DA neurons in control rats (Pucak and Grace, 1991a), it will immediately reverse depolarization block in rats treated chronically with chlorpromazine (Chiodo and Bunney, 1983). Striatonigral afferents can also be blocked chemically. Both intrinsic GABAergic collaterals (Nisenbaum et al., 1991) and striatonigral fibers (Floran et al., 1988; Giralt et al., 1990) appear to be under potent inhibitory control by $GABA_B$ presynaptic receptors. Therefore, by inhibiting neurotransmitter release from striatal cell axon terminals, systemic administration of baclofen should mimic the effects produced by a striatonigral lesion. Furthermore, although baclofen administered alone at doses up to 3 mg/kg has little effect on SN DA neuron firing in control rats, after chronic haloperidol administration a comparatively low dose of baclofen (0.1 mg/kg i.v.) reverses neuroleptic-induced depolarization block and a moderate dose (0.5 mg/kg) completely inhibits the firing of most SN DA neurons (Grace and Bunney, 1980b). Thus, although the basal activity state of DA neurons in control rats appears to be little influenced by striatonigral fibers, after chronic haloperidol administration there is a shift in the mode of regulation of DA neuron activity, with these same pathways now constituting the principle source of control over their firing.

Although the striatonigral projection appears to be required for the induction of depolarization block, haloperidol may also have a direct effect on DA cell firing by blocking DA autoreceptors located on DA neuron somata. These autoreceptors are believed to modulate DA cell activity (Groves et al., 1975) in response to DA released from their dendrites (Korf et al., 1976; Geffen et al., 1976). However, whether this putative local inhibitory system is involved in the response of DA cells to haloperidol is unclear. On the one hand, several earlier studies provided evidence that autoreceptors do not tonically inhibit DA cell firing: although the iontophoretic administration of trifluoperazine does not affect DA neuron firing (Aghajanian and Bunney, 1977), infusion of haloperidol into the striatum has been reported to increase SN DA cell firing rate (Iwatsubo and Clouet, 1977), and kainic acid lesions of the striatum were reported to block the effects of haloperidol on DA cell firing (Kondo and Iwatsubo, 1980). On the other hand, there is now substantial evidence that neuroleptics can exert direct effects on DA cells: pressure ejection of large amounts of haloperidol into the SN is reported to increase DA cell firing rate (Wuerthele et al., 1981), haloperidol has been shown to retain its ability to increase the firing rate of DA neurons after acute, carefully placed transections of the brain between the striatum and the SN (Pucak and Grace, 1991b), and DA cell inhibition by dendritic DA release can be demonstrated in SN slices recorded *in vitro* (Mercuri et al., 1990). Therefore, it appears that some of the direct excitatory effects of acutely administered haloperidol on DA cell firing can be dissociated from the induction of depolarization block by chronic neuroleptic treatment.

In contrast to the importance of the striatonigral pathway for the induction and maintenance of depolarization block in SN DA cells, the involvement of the accumbens in enabling depolarization block of VTA DA neurons has not been resolved. Although White and Wang (1983a) reported that ibotenic acid lesions of the accumbens prevented chronic haloperidol administration from inducing VTA DA cell depolarization block, Chiodo and Bunney (1983) found that, following chronic chlorpromazine administration, acute transection of the connections between the accumbens and the VTA failed to reverse DA cell depolarization block. These differences, however, may suggest that the induction of depolarization block in DA cells in the VTA is regulated separately from the maintenance of depolarization block, although nonspecific effects caused by the interruption of other fibers by the transection or alterations in DA cell regulation occurring during recovery from kainic acid lesions cannot be ruled out.

A critical appraisal of neuroleptic-induced depolarization block

Neuroleptic-induced depolarization block is a comparatively robust phenomenon that has been demonstrated electrophysiologically in a number of laboratories (e.g., Bunney and Grace, 1978; White and Wang, 1983a; Chiodo and Bunney, 1983; Meltzer et al., 1989; Kabzinski et al., 1987; Skarsfeldt, 1988; Goldstein and Litwin, 1988; Sorensen et al., 1989). Moreover, correlates of depolarization block have been described in behavioral (Rompré and Wise, 1989; Doherty and Gratton, 1991) and biochemical (Deutch and Roth, 1988; Finlay et al., 1987; Blaha and Lane, 1987; Chen et al., 1991) studies of neuroleptic action.

As discussed above, several lines of evidence support a role for neuroleptic-induced depolarization block in the therapeutic effects of antipsychotic drugs, including: 1) the similarity in time course of neuroleptic-induced depolarization block and the delayed onset of therapeutic actions, and 2) the correlation between the clinical profile of classical versus atypical antipsychotic drugs and their selective effects in eliciting depolarization block in the VTA versus the SN. Additional evidence supporting the involvement of depolarization block in antipsychotic drug actions can be drawn from a comparison of animal and clinical studies. Thus, upon discontinuation of antipsychotic drug treatment, schizophrenics showing a large increase in DA turnover are most likely to show a relapse (Glazer et al., 1989), which may be analogous to the increased DA levels occurring in rats in which depolarization block is reversed acutely by administering apomorphine (Lane and Blaha, 1987; Chen et al., 1991). Furthermore, the time course for induction of depolarization block is not affected by administering higher doses of neuroleptics (Jiang et al., 1988), which is analogous to the similarity in the time course and therapeutic efficacy associated with the use of moderate versus high doses of antipsychotic drugs in the treatment of schizophrenia (Tupin, 1985; Baldessarini et al., 1988).

Despite the impressive amount of evidence implicating depolarization block in antipsychotic drug actions, several investigators have questioned, based on biochemical measures, whether depolarization block actually occurs in DA cells in response to neuroleptic treatment. One study in particular presents biochemical evidence that purportedly refutes the presence of electrophysiologically-defined depolarization block (Andén et al., 1988). This presumption is based on the difference in effects on tissue DA levels produced in response to blocking DA cell firing by chronic neuroleptics versus the inhibition of firing elicited by systemic administration of γ-butyrolactone (GBL). However, the interpretation of these results are clouded by several factors. One shortcoming involves their reliance on the specificity of systemically administered GBL to affect the

DA system exclusively via a selective blockade of DA cell firing, which is inconsistent with current evidence regarding the actions of this drug (e.g., Hechler et al., 1991; Sneed and Bearden, 1980; Waldmeier, 1991). Indeed, the ability of GBL to decrease potassium-evoked DA release from striatal slices (Bustos and Roth, 1972) demonstrates that, in addition to blocking DA neuron firing, GBL also blocks DA release by acting directly on the DA terminal. Also, the validity of using postmortem tissue levels of DA as an index of DA system activity is a subject of contention (Kuhr et al., 1985). Furthermore, the suggestion cited by these and other investigators that the discrepancy results from the use of anesthetics in the electrophysiological studies is unclear, since neuroleptics have been shown to produce greater excitatory effects on DA cell firing in non-anesthetized rats (Bunney et al., 1973; Mereu et al., 1983). Nonetheless, the relevance of anesthetics to depolarization block has already been addressed in several experiments, which confirm the presence of chronic neuroleptic-induced depolarization block of DA neurons using immobilized, locally anesthetized rats (Bunney and Grace, 1978; Chiodo and Bunney, 1983, 1985).

The effects of depolarization block on dopamine release and metabolism

One attractive feature of depolarization block was its ability to account for the time course of the biochemical changes occurring in the DA system during repeated neuroleptic treatment — i.e., acute administration of neuroleptics causes an increase in DA cell firing rate leading to an increase in DA metabolite levels, whereas chronic treatment causes depolarization block of DA cell firing and consequently a decrease in DA metabolite levels. However, recent data suggest that this may not be an accurate representation of the phenomenon. Thus, neuroleptics are reported to increase DA metabolites in striatal slices (Iversen et al., 1976) or after interruption of nigrostriatal fibers (Garcia-Munoz et al., 1977; DiChiara et al., 1977), even though each preparation clearly circumvents any influence of DA neuron firing on these measures. Furthermore, kainic acid-induced lesions of the striatum, which are known to prevent the development of DA cell depolarization block (Bunney and Grace, 1978, 1979), do not alter the emergence of tolerance during chronic antipsychotic drug administration (Bannon et al., 1980; Biggio et al., 1980; DiChiara et al., 1977). Indeed, not only are DA metabolite levels a poor indicator of the electrophysiological activity of DA cells, it now appears that this measure is also not related to behavioral indices of DA system tolerance. Thus, the timing of the appearance of autoreceptors during development (Shalaby and Spear, 1980) corresponds closely with the age at which rats are capable of developing tolerance in DA turnover during chronic neuroleptic treatment (Coyle et al., 1985), but is nonetheless quite distinct in its developmental onset from the more functionally significant development of behavioral tolerance to neuroleptics (Coyle et al., 1985). An alternate explanation for the tolerance in DA metabolites is derived from studies suggesting that DA metabolite levels are a reflection of the rate of DA synthesis and not spike-dependent DA release (Imperato and DiChiara, 1985; Zetterstrom et al., 1988). Therefore, acute administration of antipsychotic drugs is believed to increase DA metabolites by blocking DA-mediated inhibition of synthesis-modulating autoreceptors on DA terminals (Kehr et al., 1972), leading to increased intraneuronal metabolism of DA and efflux of the DA metabolite. The tolerance in DA metabolites produced during chronic neuroleptic treatment, on the other hand, is thought to result from the induction of supersensitivity of these synthesis-modulating autoreceptors to a sufficient extent that they again fall under the modulatory influence of extracellular DA. The correlation between the development of tolerance and the induction of autoreceptor supersensitivity is supported by studies revealing their similarity in time course (Guidotti et al., 1978; Nowycky and Roth, 1977) and by the close correspondence between brain regions that fail to develop tolerance to neuroleptics and DA projection sites reported to lack synthesis-modulating autoreceptors (Bannon et al., 1981, 1982).

Because of the potential importance of understanding depolarization block and how it relates to DA release, the problems associated with interpreting the significance of DA metabolite levels were circumvented by adopting procedures that assessed brain DA levels directly using *in vivo* voltammetry or brain microdialysis techniques. Investigators have consistently shown that acute administration of classical antipsychotic drugs results in an increase in extracellular DA levels (Blaha and Lane, 1987; Zetterström et al., 1984; Imperato and DiChiara, 1985), whereas repeated administration of these drugs decreases DA levels in the striatum and accumbens (Lane and Blaha, 1987; Ichikawa and Meltzer, 1991; Finlay et al., 1987). In contrast, repeated treatment with atypical neuroleptics does not alter DA levels in the striatum (Ichikawa and Meltzer, 1991; Blaha and Lane, 1987; Chen et al., 1991; Invernizzi et al., 1990). The effects reported for chronic clozapine administration on DA levels in the accumbens, on the other hand, are not as consistent, in that studies employing voltammetry report a decrease in DA levels (Chen et al., 1991; Blaha and Lane, 1987), whereas studies employing microdialysis find no change in DA levels (Ichikawa and Meltzer, 1991; Invernizzi et al., 1990). Furthermore, only in the voltammetry studies did administration of apomorphine restore DA to control levels or above in the striatum (Lane and Blaha, 1987; Blaha and Lane, 1987) and the accumbens (Chen et al., 1991; Blaha and Lane, 1987), which is consistent with the hyperactivity of DA cells present after acute reversal of depolarization block (Bunney and Grace, 1978, 1979; Grace and Bunney, 1980b; White and Wang, 1983a; Chiodo and Bunney, 1983).

This evidence suggests that repeated neuroleptic administration attenuates spike-dependent synaptic DA release by inducing depolarization block of DA cell activity. However, this model does not provide an explanation for the source of the large residual levels of extracellular DA that are maintained in chronically treated rats despite the inactivation of

DA cell firing. Unfortunately, correlating extracellular DA with spike firing in rats after chronic drug treatment introduces unknown variables that cloud the interpretation of these data. For example, it is unclear whether DA levels are initially depressed by depolarization block but recover over time via the augmentation of some type of spike-independent release, or if there is a dissociation between DA cell firing and extracellular DA levels. Indeed, the dependence of extracellular DA levels on synaptic DA release is not clear. A further complication may be derived from methodological shortcomings regarding the techniques used to measure DA levels. For example, the discrepancy between the voltammetry studies and the dialysis work may point to a potential source of artifact: since striatal lesions or striatonigral transections are reported to reverse depolarization block (Bunney and Grace, 1978, 1979; White and Wang, 1983a), one must be concerned about the possibility that implantation of the rather massive dialysis probe into the striatum may cause sufficient anatomical or neurochemical disruption (O'Neill et al., 1991) to reverse DA neuron depolarization block. In order to resolve these questions, a method of inducing depolarization block acutely after the implantation of the dialysis probes was developed.

Assessing the effects of depolarization block using an acute model

As outlined previously, induction of depolarization block by neuroleptics requires long-term treatment, since acute administration of haloperidol will not induce depolarization block in DA neurons even when extremely high doses are used (Hollerman and Grace, 1989a,b). The time course of this phenomenon, moreover, appears to be rather circumscribed, in that depolarization block requires approximately the same amount of time to develop regardless of whether moderate or high doses of haloperidol are administered (Jiang et al., 1988). A time-dependent change in the response of an organism during repeated drug administration may imply that a modification in receptor sensitivity is occurring (Friedhoff and Miller, 1983), and thus may also account for the similarity in the time course of what appear to be etiologically independent responses occurring during chronic drug treatment, such as the Parkinsonian side effects, tolerance in DA metabolite levels, and antipsychotic efficacy. Although the increase in autoreceptor sensitivity that occurs during chronic treatment appears to account for the development of tolerance in DA metabolites (Bannon et al., 1982; Guidotti et al., 1978), it is unlikely that a change in either autoreceptor or postsynaptic DA receptor sensitivity can account for depolarization block, since administration of near lethal doses of haloperidol should saturate the DA receptors regardless of their state of sensitivity. Thus, the locus of receptor change is likely to occur at a non-DA synapse located between the striatal dopaminoceptive neuron and the SN DA cell. A likely candidate would be at GABAergic synapses that comprise a major portion of this projection system. This is consistent with the alterations in GABA sensitivity reported to occur along this pathway during chronic treatment (Frey et al., 1987, 1989; Gale, 1980). In this context, neuroleptic administration is not the only means by which striatonigral GABAergic systems are affected. In fact, qualitatively similar changes in GABAergic sensitivity have been observed when DA transmission is compromised through another means — i.e., by partial lesions of the nigrostriatal DA fibers (Pan et al., 1985; Weick et al., 1990). Rats which have initially been made akinetic as a result of partial DA-depleting lesions have also been shown to undergo a remarkable degree of behavioral recovery, regaining basal motor function within 4–6 weeks following the lesions (Zigmond and Stricker, 1973; Zigmond et al., 1990). However, despite their normal appearance, placing additional demand on the DA system by administering neuroleptics in doses too low to affect control rats will rapidly reinstate motor deficits (Heffner et al., 1977; Hollerman and Grace, 1989b; Hollerman et al., 1992; Snyder et al., 1985). Using single unit recording techniques, we examined whether the increased behavioral sensitivity to haloperidol was correlated with changes in the electrophysiological response of DA cells to this neuroleptic. Consistent with the findings of others, haloperidol only increased the firing rate of DA neurons in intact rats, even when near lethal doses of the drug were given. Similarly, in rats with 80%–90% DA depletions tested within one week after the lesion (i.e., before many of the compensatory processes had taken place; Zigmond et al., 1990), very large doses of haloperidol (3–4 mg/kg i.v.) again failed to induce depolarization block in the remaining DA neurons. However, in rats that had recovered behaviorally by one month after the DA-depleting lesions, very low doses of haloperidol (0.1 mg/kg i.v.) caused most of the remaining DA neurons to immediately enter a state of depolarization block (Hollerman and Grace, 1989a,b; Hollerman et al., 1992). Indeed, the dependence of acute neuroleptic-induced depolarization block on the degree of DA depletion present in the rat may account for the reported wide variation between individual patients in the time course of their therapeutic response and onset of parkinsonian side effects during antipsychotic drug treatment.

The ability to induce depolarization block acutely in this preparation permitted us to assess the immediate effect of depolarization block on DA levels in the striatum using *in vivo* microdialysis measures. Administration of haloperidol (0.5 mg/kg i.p.) to control rats was found to increase extracellular DA levels in the striatum by 95% over baseline. In contrast, in the lesioned/recovered animals administration of the same dose of haloperidol, which should have been sufficient to drive most of the residual DA neurons into depolarization block, did not produce a change in extracellular DA levels (Abercrombie et al., 1989; Hollerman et al., 1992). It should be noted that this result was not due to a "ceiling effect," since stress was still capable of increasing DA levels in the striatum of rats with similar lesions (Keefe et al., 1990).

The ability of neuroleptics to inactivate DA neuron firing by depolarization block instead of by hyperpolarization may be an important factor in their therapeutic actions. Thus, a unique feature of DA neurons in depolarization block is that they cannot increase spike-dependent DA release in response to stimuli that normally are excitatory. This interpretation is consistent with the results of studies by Kamata and coworkers (1986), which show that the increase in striatal DA levels elicited in control rats by GABA agonist-mediated disinhibition of DA neurons cannot be produced in rats following neuroleptic-induced inactivation of DA neuron firing. This mode of therapeutic action also is consistent with what we know about the plasticity of this system — i.e., that quantitative changes in the absolute levels of neurotransmitter are relatively unimportant since they can be counteracted via homeostatic compensations, whereas producing a qualitative change in the way a system responds to stimuli is more likely to induce long-term alterations in its function. Nonetheless, despite the apparent functional relevance of depolarization block to the DA system, it still does not provide an explanation as to why the induction of depolarization block is not accompanied by a profound decrease in extracellular DA levels. Furthermore, if schizophrenia can be treated by changing how DA cells respond to stimuli, what could be the nature of the neurophysiological dysfunction that leads to schizophrenia in the first place? These important questions may be most effectively addressed by first examining the relevance of extracellular DA levels to the physiology of this system.

A model of schizophrenia based on the therapeutic relevance of neuroleptic-induced depolarization block

In order to assess the impact of depolarization block on DA release, one must determine whether extracellular DA levels or DA metabolites are indeed a valid measure of DA released by impulse flow. In fact, most evidence suggests that these two variables are unrelated (Pani et al., 1990; Grace, 1991). Extracellular DA levels do not reflect DA release directly, but instead represent DA *overflow* — i.e., the amount of DA that escapes the synaptic cleft after release (Ungerstedt et al., 1982; Zetterström et al., 1983). In actuality, the vast majority of the DA that is released into the synaptic cleft apparently is taken back up into the DA terminal. Using fast-scan voltammetry, Wightman and colleagues have found that brief periods of stimulation delivered to DA cell axons could produce short-term increases in DA levels in the extracellular space of the striatum only if the DA cell axons were stimulated at frequencies of 10–20 Hz or greater (Kuhr et al., 1984; Kuhr and Wightman, 1986; May et al., 1988), which is 2–3 times faster than DA cells are capable of firing (Grace and Bunney, 1986). Moreover, even with these supramaximal stimulation frequencies, the resultant increase in DA levels was not maintained for more than a few seconds (Keller et al., 1988; Ewing and Wightman, 1984). Although some laboratories using the more sensitive microdialysis technique have on occasion reported small but measurable changes in DA levels using more

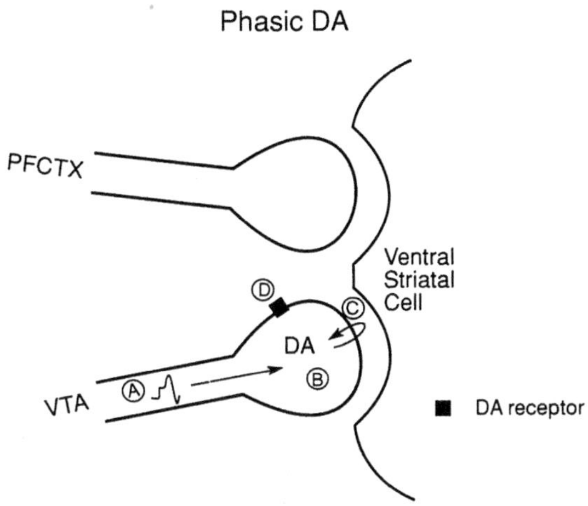

Fig. 1. A model of the phasic DA response system. Environmental stimuli activate VTA DA cell spike firing (*A*). Upon invading the DA terminal in the ventral striatum (*B*), the spikes trigger DA efflux into the synaptic cleft (*C*), where it stimulates DA receptors on postsynaptic cell spines. Soon after its release, the DA is rapidly cleared from the synaptic cleft by uptake into the DA terminal (*C*) before it can escape into the extrasynaptic space and stimulate DA nerve terminal autoreceptors (*D*) or activate other homeostatic processes. In this model, stimulation of postsynaptic DA receptors by this spike-dependent DA release is proposed to underlie the behavioral actions of this neurotransmitter. Figures 1–5 have been adapted from the model described in Grace (1991). Abbreviations used in Figs. 1–5: *DA* dopamine; *Glu* glutamate; *PFCTX* prefrontal cortex; *VTA* ventral tegmental area

physiological rates of stimulation, this occurs only when the stimulation is presented continuously for 10–20 minutes or longer. Therefore, stimulating the DA terminal at physiologically relevant frequencies will at best elicit only a transient change in extracellular DA levels. Nonetheless, it is clear that synaptic DA release does occur in response to physiologically relevant rates of stimulation. For example, DA-dependent behavioral effects can be elicited in rats when DA cell axons are stimulated at only 4–5 Hz even though no measurable change in extracellular DA levels is produced (Fibiger et al., 1987; Gratton et al., 1988; Nakahara et al., 1989). Therefore, it appears that physiologically relevant rates of DA axon discharge will produce behaviorally effective increases in synaptic DA levels, with the fast reuptake process rapidly removing the DA from the synaptic cleft before it can reach the extracellular space (May et al., 1988). Because of the brief nature of this event, this type of DA action has been termed the *phasic* DA response (Fig. 1).

If DA that is released in response to spike discharge is inactivated by reuptake before it can leave the synaptic space, it is unlikely that the constant levels of extracellular DA that can be measured in the striatum are produced by DA neuron firing. Indeed, several investigators have proposed that DA levels, as measured by dialysis, voltammetry, or push-pull cannula techniques, are regulated primarily by local factors, with DA cell discharge playing a minor role. This is based on studies showing that direct presynaptic activation of DA terminals by glutamate administration (Glowinski et al., 1988), stimulation of corticostriatal afferents (Nieoullon et al., 1978), or cortical activation via the thalamus (Romo et al., 1986; Chesselet et al., 1983), will increase DA levels in the striatum *in vivo* without producing a concomitant activation of DA cell firing. Furthermore, the change in DA levels produced is small in amplitude and occurs with a very slow time course. In many respects, this DA release appears to be analogous to the TTX-sensitive, calcium-dependent NMDA-mediated DA release reported by others (Clow and Jhamandas, 1989). DA released in this manner apparently can be regulated independent of the TTX-sensitive DA release produced by DA cell discharge, although evidence for the involvement of non-NMDA-linked glutamate receptors in DA release has been advanced recently (Imperato et al., 1990; Barbieto et al., 1990). Furthermore, this heterosynaptic afferent-mediated DA release may have a functional role in the response of this system to stress, since stress can induce increases in DA metabolites that can be blocked by NMDA antagonists (Serrano et al., 1989) without activating DA cell firing (Strecker and Jacobs, 1985). Its independence from spike-induced DA release is further substantiated in studies showing that repeated administration of neuroleptics at doses sufficient to cause depolarization block of DA cell firing does not block the increase in DA metabolite levels caused by stress (Davila et al., 1988b). Although recent experiments have cited the need for high glutamate concentrations to increase DA levels and the failure of glutamate antagonists to decrease basal DA release (Moghaddam et al., 1990; Moghaddam and Gruen, 1991) as reasons to question the specificity of glutamate-induced DA release, these studies fail to consider how glutamate-induced activation of intrinsic GABAergic neurons within the striatum (Spencer, 1976; Herrling, 1985) may affect their results. The fact that GABA can decrease DA release (Reimann et al., 1982) and that GABA antagonists potentiate the effects of glutamate on DA release (Chéramy et al., 1986; Leviel et al., 1990) suggest that the presynaptic release of DA by exogenous glutamate may be counterbalanced by glutamate-induced activation of striatal GABAergic neuron firing.

Tonic DA levels determine the "background" level of DA receptor stimulation, and therefore are closely regulated by homeostatic processes. Thus, one month after depletions of up to 80% of tissue DA levels, the extracellular DA levels in the striatum return to control values (Robinson and Whishaw, 1988). When the lesion size is greater than 80% there is a decrease in extracellular DA concentration; nonetheless, the baseline level of receptor stimulation is maintained at control levels (Nisenbaum et al., 1986) by a concomitant increase in DA receptor sensitivity, which also occurs only when DA depletions exceed 80% (Mishra et al., 1980). Thus, the homeostatic regulatory system appears to function primarily by restoring this background level of DA receptor stimulation. Due to the

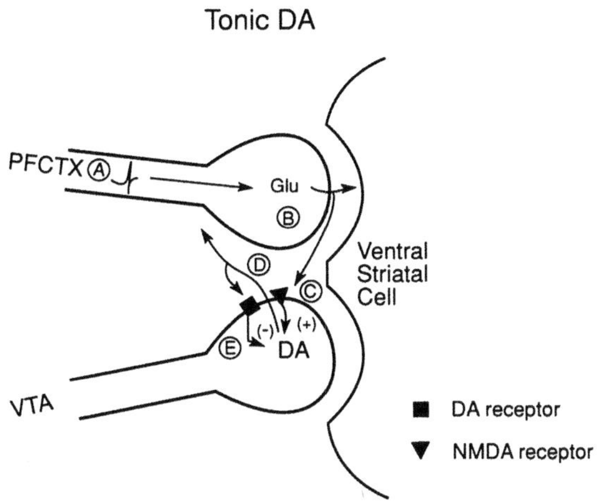

Fig. 2. A model of the tonic DA response system. Spike activity in the prefrontal corticostriatal fibers (*A*) triggers glutamate release from terminals in the ventral striatum (*B*), where it stimulates glutamate receptors on postsynaptic cell spines. In addition, a portion of the glutamate escapes from the synaptic cleft to augment glutamate levels in the extracellular space. Glutamate levels in the vicinity of the DA terminal are sufficient to provide tonic stimulation of the sensitive NMDA receptor on DA terminals (*C*), evoking spike-independent DA release into the extracellar space (*D*). Despite the comparatively low concentration of DA in the extracellular space, it is nonetheless adequate to provide a constant level of DA terminal autoreceptor stimulation, which causes feedback inhibition of DA synthesis and release (*E*)

constant, low amplitude DA receptor stimulation provided by this system, it has been called the *tonic* DA response (Fig. 2). Of course, since the extracellular levels of DA in the striatum are so low, one may question whether tonic concentrations of DA have any physiological significance. However, a functional role for tonic DA and its regulation by glutamate may be predicted based on the following considerations: (1) The extracellular level of glutamate in the striatum (2–8 µM; Korf and Venema, 1985) is sufficient to tonically activate NMDA receptors (ED_{50} = 5 µM; Sands and Barish, 1989) as shown to occur in the hippocampus (Sah et al., 1989). (2) The extracellular level of DA in the striatum (10–50 nM; Church et al., 1987; Sharp et al., 1986) should be sufficient to tonically activate D_2 receptors (K_d = 40 nM; Richfield et al., 1989) which, among other functions, will presynaptically inhibit spike-dependent DA release (Farnebo and Hamberger, 1971). (3) Each striatal neuron is estimated to receive 4,500–8,000 DA terminals, with half of these synapsing on the same dendritic spine as the corticostriatal afferents (Doucet et al., 1986), thus supplying each neuron with locally high concentrations of extracellular DA and glutamate. Indeed, it may be necessary to modify our view of the pharmacological regulation of the striatal neuron to include two neurochemical processes: one mediated by direct synaptic connections releasing high concentrations of neurotransmitter into the synaptic cleft, super-

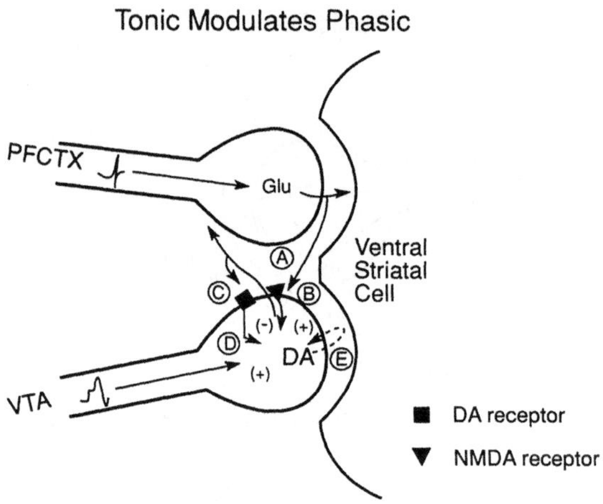

Fig. 3. Influence of tonic DA release on the amplitude of the phasic DA response. Activity in corticostriatal afferents elicits glutamate release (*A*) that augments extracellular glutamate levels in the ventral striatum. The concentration of glutamate in the vicinity of the DA terminal provides a constant level of glutamate stimulation to the DA terminal (*B*), evoking NMDA-dependent DA release into the extracellular space (*C*). DA levels in the vicinity of the DA terminal stimulate DA autoreceptors (*D*) which provide feedback inhibition of spike-mediated phasic DA release. In contrast, the larger amplitude, spike-dependent DA release into the synaptic cleft is sufficient to trigger the less sensitive postsynaptic DA receptors on the striatal neuron, and then is rapidly taken up into the DA terminal before it can escape the synaptic cleft (*E*). In this way, cortically-mediated release of DA into the extracellular space serves to attenuate spike-dependent DA release and thereby decrease stimulation of postsynaptic DA receptors

imposed upon a "cloud" of neurotransmitters bathing the neuron and its afferent synaptic terminals with comparatively low concentrations of transmitter.

In contrast to the comparatively stable levels of tonic DA in the extracellular space of the striatum, DA that is released in a phasic manner appears to be eliminated too quickly to allow it to activate these homeostatic processes. However, since tonic DA levels are sufficient to stimulate DA terminal autoreceptors, which will down-regulate the impulse-dependent phasic DA release (Farnebo and Hamberger, 1971), any changes in tonic DA levels should be capable of modulating the amplitude of the phasic DA response (Fig. 3). Thus, cortically-mediated increases in tonic DA levels would be expected to functionally oppose stimuli that cause phasic DA release by activating DA neuron firing. Indeed, experimental evidence shows that the extracellular level of DA in the striatum is sufficient to decrease stimulus-induced DA release from DA terminals (Reimann et al., 1979).

The potential interactions of these systems may be of particular relevance when one considers the changes that are believed to occur in schizophrenics. Several lines of investigation have provided evidence for decreased activity in the frontal cortex of schizophrenics (i.e., hypofrontality): (1) the frontal cortex of schizophrenics does not show increased cerebral blood flow when the patient is presented with a specific discrimination task (Ingvar and Franzen, 1974; Weinberger et al., 1986), (2) studies using magnetic resonance spectroscopy show that schizophrenics have decreased metabolic activity in their frontal cortices somewhat analogous to premature aging (Pettegrew et al., 1989, 1991; Williamson et al., 1991), and (3) anatomical studies reveal evidence of pathological changes in the frontal cortex in schizophrenics (Shelton et al., 1988). Since cortical pyramidal cell projections are typically glutamatergic (Spencer, 1976), hypofrontality should be accompanied by diminished activity of the glutamate system innervating subcortical structures. This is consistent with several studies reporting diminished glutamate function in schizophrenics, including a decrease in CSF levels of glutamate (Kim et al., 1980; but see Perry, 1982), an increase in glutamate binding in the striatum (Kornhuber et al., 1989), increased kainate binding in the frontal cortex (presumably to presynaptic autoreceptors on glutamatergic terminals; Nishikawa et al., 1983a; Deakin et al., 1989), and a diminished efflux of glutamate from synaptosomes derived from schizophrenic brains (Sherman et al., 1991). A decrease in glutamate release in the ventral striatum should, in turn, decrease glutamate-stimulated tonic DA release, which is consistent with the reported negative correlation between glutamate levels and HVA concentration in the CSF of schizophrenics (Alfredsson et al., 1988).

The proposed relationship between cortical function and DA release is supported by studies showing that schizophrenics with cortical damage have significantly lower levels of DA metabolites than the normal population (Van Kammen et al., 1983; Nyback et al., 1983). If a decrease in glutamate release from corticostriatal afferents translates into a decrease in glutamate-stimulated tonic DA release in the ventral striatum of schizophrenics, one result would be an abnormal up-regulation of spike-dependent phasic DA release (Fig. 4). Thus, any stimuli that exert part of their behavioral actions via an activation of DA neuron firing would cause an abnormally large level of phasic DA release. The abnormally large phasic DA release is proposed to underlie the positive symptoms of schizophrenia, which appear to be most closely related to a functional hyperdopaminergic state: (1) only the positive symptoms are exacerbated in schizophrenics by amphetamine (Angrist et al., 1980), and (2) these symptoms respond to classical antipsychotic drug treatment (Pickar et al., 1984, 1986). According to this perspective, neuroleptic-induced depolarization block would not alter the tonic level of DA in the ventral striatum, but should nonetheless circumvent stimulus-induced increases in phasic DA release by preventing activation of DA neuron firing (Fig. 5).

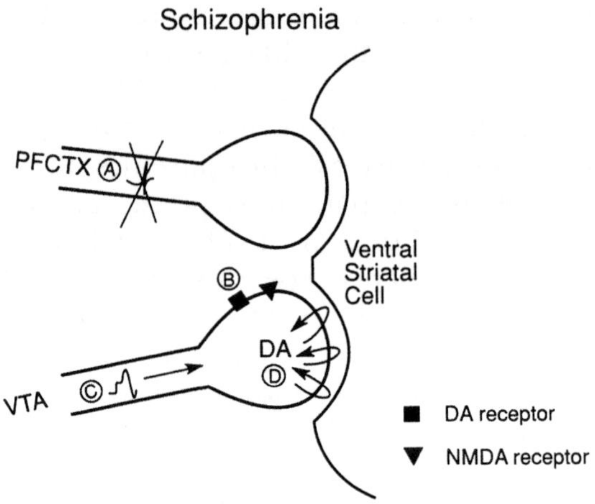

Fig. 4. Evidence from several lines of investigation suggest that schizophrenics may have deficits in cortical glutamatergic systems. In the model of schizophrenia outlined here, a pathological decline in corticostriatal glutamate release (*A*) causes a decline in glutamate-dependent tonic DA levels. The resultant decrease in DA autoreceptor stimulation would remove feedback inhibition of DA synthesis and release (*B*). In this state, environmental stimuli that activate DA neuron firing (*C*) would evoke an abnormally large spike-dependent DA release from DA terminals (*D*) in the ventral striatum. This abnormal potentiation of stimulus-evoked phasic DA responses is proposed to underlie the positive symptoms of schizophrenia

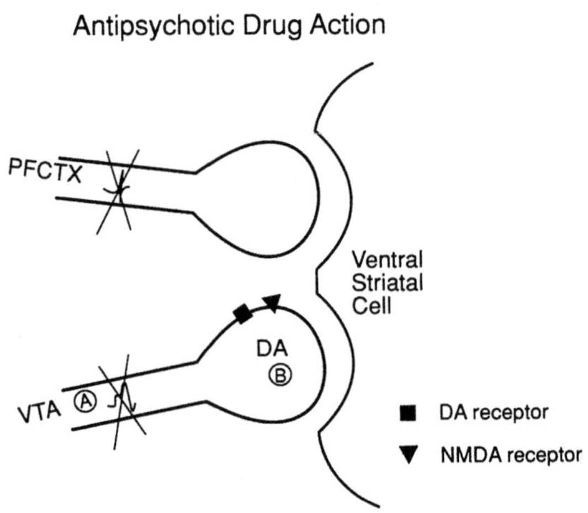

Fig. 5. Although antipsychotic drugs block DA receptors within minutes after acute administration, they must be administered repeatedly over a period of weeks in order to achieve significant therapeutic benefits. This delayed action corresponds to the time course of antipsychotic drug-induced depolarization block of DA cell firing. By preventing stimuli from activating DA cell firing, neuroleptic-induced depolarization block (*A*) is proposed to prevent spike-dependent DA release (*B*), thereby circumventing the abnormally large phasic DA response in the ventral striatum

Schizophrenia as a deficiency of dopamine system regulation: speculations and parallels with other disease states

One implication of this model of psychotic disorders, which has been described in full elsewhere (Grace, 1991), is that the primary pathological disturbance in schizophrenia is not within the DA system; instead, the disorder is proposed to result from an abnormal cortical regulation of an otherwise normal DA system. Furthermore, although the frontal cortex is described in this model as the site of pathological alteration, deficits in several brain regions could yield equivalent results. In fact, most brain regions reported to show pathological changes in schizophrenics, such as the hippocampus (Bogerts et al., 1985; Conrad et al., 1991), amygdala (Tamminga et al., 1991; Reivich et al., 1979), cingulate cortex (Benes and Bird, 1987), and limbic and frontal cortices (Roberts and Bruton, 1990; Shelton et al., 1988), also have glutamatergic projections to the limbic striatum (McGeorge and Faull, 1989). It is possible that the different types of psychoses observed in humans may be related to the cortical site where the primary pathological disturbance is located. This would account for the ability of antipsychotic drugs to treat all types of psychoses (Baldessarini, 1985; Friedhoff and Miller, 1983), since each would involve a final common action on the subcortical DA system.

According to this perspective, the therapeutic actions produced by chronic antipsychotic drug-induced depolarization block are not dependent on a decrease in the absolute level of DA, since such a change could be readily circumvented by the homeostatic mechanisms that regulate this system. Instead, depolarization block of DA cell spike generation would serve to suppress the abnormal phasic DA response by preventing excitatory stimuli from increasing DA cell firing. Interpreted in this way, the debate over the discrepant findings regarding the effects of chronic clozapine administration on extracellular DA levels in the accumbens becomes moot; that is, the therapeutically relevant effect of neuroleptics (at least with respect to the positive symptoms) would occur via depolarization block of spike-dependent synaptic DA release independent of their actions on extracellular DA concentrations. Nonetheless, one must consider whether the unique ability of clozapine to increase tonic extracellular DA levels in subcortical structures (Ichikawa and Meltzer, 1991) may play a role in its clinical actions. Although the relevance of extracellular DA to the positive symptoms of schizophrenia is unclear, this observation may provide insight into why clozapine may be particularly effective in alleviating the negative symptoms of schizophrenia (Kane et al., 1988; Meltzer, 1989). Prominent negative symptoms are often associated with: (1) a higher probability of damage to the cortex, amygdala, or hippocampus (Alpert et al., 1978; Meltzer, 1984; Van Kammen et al., 1986; Bogerts, 1985), (2) lower initial levels of HVA (Lindstrom, 1985; Van Kammen et al., 1983), (3) attenuated HVA response to acute neuroleptic administration (Bowers et al., 1984, 1989; Van Kammen et al., 1986), and (4) diminished therapeutic response to treatment with classical antipsychotic drugs (Brown and Herz, 1989). However, one type of treatment that appears to be effective in controlling negative symptoms is the administration of DA agonists, such as apomorphine (Tamminga et al., 1978), L-DOPA (Gerlach and Luhdorf, 1975; Meltzer et al., 1986), or amphetamine (Angrist et al., 1980). One feature that chronic clozapine administration and DA agonist treatment have in common is their ability to increase tonic DA receptor stimulation. Therefore, in this model, negative symptoms would be produced in schizophrenics who had sufficient cortical damage to cause a significant loss of extracellular DA levels in subcortical structures, with clozapine being effective in treating negative symptoms as a consequence of its ability to increase tonic DA levels. Under these circumstances, it clearly would be premature to attempt to correlate the entire spectrum of clinical efficacy of a drug to a single attribute, such as its ability to induce depolarization block (Meltzer, 1991).

This model is also consistent with what is known about the actions of phencyclidine (PCP) in the brain. PCP is reported to produce a state in normals that more closely models schizophrenia than does amphetamine-induced psychosis because, although both drugs produce positive symptoms (Allen and Young, 1978; Connell, 1958), only PCP is reported to elicit negative signs as well (Javitt, 1987). Similarly, although both amphetamine and PCP increase spike-dependent DA release and block DA reuptake (Taylor and Snyder, 1971; Vickroy and Johnson, 1983), PCP exerts additional pharmacological actions through its ability to decrease NMDA receptor activation (Snell and Johnson, 1985). According to the present model, a PCP-induced decrease in cortically-mediated tonic DA release would be expected to produce symptoms resembling the negative signs of schizophrenia (Grace, 1991).

It is important to note that although neuroleptics are effective at controlling some of the symptoms of schizophrenia, it is likely that they are *not* doing so by specifically reversing the core cause of the disorder, since: (1) neuroleptics exert their antipsychotic activities without reversing the hypofrontality (Buchsbaum et al., 1987) or metabolic deficits (Pettegrew et al., 1989, 1991) that may be involved in the etiology of this disorder, and (2) it is likely that the depolarization block induced by neuroleptics does not represent a restoration of the "normal" state of DA cell activity in control subjects. Instead, neuroleptics appear to exert their therapeutic actions by inducing an abnormal state of DA neuron activity which counterbalances the abnormal DA regulation produced by pathological changes in the cortex of schizophrenics. One reason that the current therapeutic regimen used in treating schizophrenics is comparatively nonspecific and prone to inducing side effects may be that our current method for screening for antipsychotic efficacy is targeted at the "symptoms" of schizophrenia (i.e., the disordered DA system regulation) rather than the causative factors (i.e., disruption of the cortical glutamatergic system). For example, the most commonly employed technique used as an initial screen for new antipsychotic drugs is based on their acute effects on rats in performing the conditioned avoidance response (Cook and Catania, 1964; Davidson and Weidle, 1976). Given that antipsychotic drugs require weeks of treatment to achieve therapeutic effects, it is unlikely that their acute behavioral actions can accurately reflect their therapeutic potential (Friedhoff and Miller, 1983). Thus, the development of new drugs and their implementation into treatment strategies employed today for schizophrenics seem to be based primarily on how closely they mimic the pharmacological actions of classical neuroleptics such as chlorpromazine, which was discovered "fortuitously" rather than by rational drug design targeted at the clinical deficit itself.

These considerations suggest that the biological bases of psychiatric disorders may be more effectively analyzed by abandoning simple models of disorders that are based on excesses or deficiencies of neurotransmitters in the brain in favor of more complex models emphasizing dysfunctions in the manner in which neurotransmitter systems are regulated. This concept is based, in part, on the recognition that the normal brain is most effective at circumventing deviations in steady-state neurotransmitter levels. For example, studies of recovery in the DA system after lesions illustrate the extensive capacity of the normal brain to compensate for this type of deficit. This is not to deny that functional deficits can be derived from absolute changes in neurotransmitter levels; in fact, it is these "steady-state" disorders, such as anxiety, epilepsy, and Parkinson's disease, for which we have the most precise animal models. This is due in part to another characteristic of steady-state disorders: they produce a measurable change in the brain, ranging from physiological disturbances in brain waves or neuronal hyperexcitability in epileptics to decreases in DA and DA metabolites in Parkinson's disease. These steady-state disorders are also distinct with respect to the actions of therapeutically effective drugs: (1) the drugs have immediate therapeutic actions when administered acutely, reversing the symptoms relatively soon after their administration, and (2) there is a progressive decrease in therapeutic efficacy as the drugs are administered repeatedly over an extended period of time. Indeed, the drugs used to treat these disorders have their peak therapeutic value the first time they are administered, with their therapeutic actions derived from the direct pharmacological properties of the drug.

This is in marked contrast to the characteristics of the "regulatory" disorders. Among the first psychiatric disorders proposed to result from a regulatory dysfunction is depression (Maas, 1979; Siever and Davis, 1985). Although the initial set of criteria proposed for disorders of this type were somewhat narrow in focussing on cyclic disorders, assessing the predicted characteristics of regulatory disorders in more general terms enables one to include schizophrenia in this definition. The regulatory disorders are proposed to have the following properties in common: (1) the disorders usually are not accompanied by quantifiable changes in brain neurotransmitter levels or gross physiological measures, (2) the drugs used in their treatment have little or no therapeutic value when administered acutely, but require repeated administration for their therapeutic benefits to be realized, and (3) the therapeutically relevant actions of these drugs are different from the effects exerted by acute administration. Thus, unlike the acutely active drugs, their delayed therapeutic actions are distinct from their acute pharmacological profile, exerting their effects by inducing changes in the way that a system is regulated or in how it responds to stimuli. For example, although the acute pharmacological profile for antipsychotic drugs is based on DA receptor blockade, their primary therapeutic effects appear to depend on preventing DA cell activation by inducing depolarization block. In fact, for many drugs in this class, the short-term side effects produced by the drugs are an extension of their acute pharmacological actions — e.g., akinesia during acute DA receptor blockade. For this reason, the homeostatic responses elicited by these drugs could be responsible for the observation that these early side effects often diminish during treatment at the same time that the long-term therapeutic actions are developing.

The ideal drug for the treatment of regulatory disorders would be one that produces highly specific changes at the site of pathology to restore the responses of the system to normal. Under such circumstances, one may not expect such drugs to exert significant actions on normal systems. Indeed, such a condition may contribute to the difficulty encountered in determining the site of action of highly effective medications such as tricyclic antidepressants or lithium — i.e., they may elicit substantial neurochemical changes only in systems that are defective to begin with, while having little or no effect on normal subjects. However, what about drugs, such as antipsychotic drugs, which have rather specific pharmacological profiles and presumably exert the same effects in schizophrenics as in normals? One possible manner in which apparent "non-regulatory" antagonists exert therapeutic actions on "regulatory" disorders could be via an indirect action that occurs as a consequence of "overwhelming" the system by receptor blockade in a manner analogous to that proposed for their association with extrapyramidal side effects (e.g., drug-induced parkinsonism, tardive dyskinesia, akathesia, etc). Thus, the limited efficacy of antipsychotic drugs may be related to the fact that they cause the same profound and clear-cut changes in the DA system of both normals and schizophrenics presumably by acting in the same manner in each subject: i.e., producing depolarization block rather than by a specific action at the etiological locus of the disorder. In contrast, drugs lacking such a clearly defined steady-state pharmacological action may exert such potent and specific effects on diseases because they act on the specific etiological factor that is responsible for the disorder. Such disease-specific drugs should produce little effect when taken by control subjects, and therefore may not be as amenable to investigation by using "normal" animals. Therefore, uncovering the precise mode of action of antidepressants, which may act by restoring a dysfunctional system to "normal" (Maas, 1979), would be greatly facilitated if one could first identify the irregularity that is being corrected in the brains of patients with affective disorders. If this model is indeed generally applicable across disease states, it may also suggest that far more effective antipsychotic drugs will be developed once we better understand the pathological disturbances that underlie schizophrenia.

Summary and conclusions

Antipsychotic drugs have proven to be very effective treatments for schizophrenia, despite their propensity for inducing side-effects. The ability of antipsychotics to block the neurotransmitter DA is, on the surface, consistent with the pharmacological evidence favoring the DA hypothesis of schizophrenia. However, this hyperdopaminergic model has been cast into doubt based on: (1) the lack of evidence for increased DA turnover in schizophrenics, (2) the fact that antipsychotic drugs block DA receptors immediately but require repeated treatment for significant therapeutic effects to occur, and (3) the pharmacological profile of the atypical antipsychotic drugs. Therefore, the ability of antipsychotic drugs to exert therapeutic actions is apparently dependent on a time-dependent change distinct from their acute DA blocking properties.

One of the first neurochemical changes that was found to correlate with the time course of antipsychotic efficacy was the development of tolerance in DA turnover during chronic antipsychotic drug treatment. At first, the development of depolarization block of DA neuron firing was proposed as the mechanism underlying this tolerance. However, it is now apparent that the development of tolerance can occur independent of depolarization block, and instead may be an epiphenomenon that is correlated with the emergence of supersensitivity in DA synthesis-modulating autoreceptors. In contrast, both the antipsychotic efficacy and capacity for inducing extrapyramidal side effects was found to correlate highly with the ability of these drugs to induce depolarization block of DA neuron firing. The correlation in the time course of the development of tolerance with the induction of depolarization block and the onset of antipsychotic efficacy instead may be explained if one assumes that in both cases the delay reflects a time-dependent change in the sensitivity of distinct receptor systems.

Although chronic administration of classic antipsychotic drugs will decrease extracellular DA levels during the course of treatment, one may predict that the high capacity of the DA system to homeostatically compensate for changes could readily reverse such a decrease. Furthermore, this antipsychotic drug-induced decrease in extracellular DA levels is not as apparent in studies examining the effects produced by repeated administration of the more efficacious atypical antipsychotic drugs. However, in many cases extracellular DA is not directly related to the electrophysiological activity of DA neurons. Instead, the therapeutically significant effect of DA neuron depolarization block is likely to involve its ability to prevent further activation of DA neuron firing by stimuli. Taking this evidence into consideration, together with more recent evidence supporting a cortical dysfunction in schizophrenics, a new model of schizophrenia was derived (Grace, 1991). This model is based on the differential regulation of DA release by two distinct processes: (1) a presynaptic glutamate-induced DA release that allows the cortex to regulate tonic extracellular DA levels, and (2) a DA neuron spike-dependent phasic release of DA into the synapse that is rapidly inactivated by reuptake immediately after its release, and thus does not affect extracellular DA levels. This model proposes that inactivation in regions of the cortex in schizophrenics causes a fall in extracellular DA levels, causing a homeostatic compensatory up-regulation of the DA system and a release of DA terminals from presynaptic inhibition. As a consequence, activation of DA neuron firing by stimuli would produce pathologically high levels of synaptic DA actions. By inactivating DA neuron firing via overexcitation, neuroleptic-induced depolarization block of DA cells would functionally oppose this effect by preventing further excitation of DA neuron firing.

According to this model, the site of pathology in the schizophrenic occurs in one or more of the many systems providing glutamate afferents to the ventral striatum, with the final common action being an abnormal up-regulation of the DA system. Thus, induction of depolarization block by neuroleptics fails to alter the specific pathological origin of the disorder, instead producing an offsetting deficit by inactivation of all spike-dependent DA release. New therapeutic strategies based on reversing this abnormal cortical regulation of the DA system may yield more effective treatments of schizophrenia that additionally have a lower capacity for inducing extra-pyramidal side effects.

Acknowledgements

I would like to thank E. M. Stricker and J. R. Hollerman for their helpful comments in the preparation of this manuscript. This work was supported by USPHS MH42217, MH45156 and NS19608.

References

Abercrombie ED, Hollerman JR, Grace AA (1989) In vivo biochemical correlates of acute depolarization inactivation in substantia nigra dopaminergic neurons. Soc Neurosci Abstr 15; 1002

Aghajanian GK, Bunney BS (1974) Dopaminergic and non-dopaminergic neurons of the substantia nigra: differential responses to putative transmitters. In: Boissier JR, Hippius H, Picho P (eds) Proceedings 9th International Congress of the Collegium Internationale Neuropsychopharmacologicum. Excerpta Medica, Amsterdam, pp 444–452

Aghajanian GK, Bunney BS (1977) Dopamine "autoreceptors": pharmacological characterization by microiontophoretic single cell recording studies. Naunyn-Schmiedebergs Arch Pharmacol 297: 1–7

Alfredsson G, Wiesel F-A, Tylec A (1988) Relationships between glutamate and monoamine metabolites in cerebrospinal fluid and serum in healthy volunteers. Biol Psychiatry 23: 689–697
Allen RM, Young SJ (1978) Phencyclidine-induced psychosis. Am J Psychiatry 135: 1081–1084
Alpert M, Friedhoff AJ, Marcos LR, Diamond F (1978) Paradoxical reactions to L-dopa in schizophrenic patients. Am J Psychiatry 135: 1327–1332
Andén N-E, Grenhoff J, Svensson TH (1988) Does treatment with haloperidol for 3 weeks produce depolarization block in midbrain dopamine neurons of unanesthetized rats? Psychopharmacology 96: 558–560
Angrist BM, Shopsin B, Gershon S (1971) The comparative psychotomimetic effects of stereoisomers of amphetamine. Nature (London) 234: 152–153
Angrist BM, Rotrosen J, Gershon S (1980) Differential effects of amphetamine and neuroleptics on negative vs. positive symptoms in schizophrenia. Psychopharmacology 72: 17–19
Baldessarini RJ (1985) Drugs and the treatment of psychiatric disorders. In: Gilman AG, Goodman LS, Rall TW, Murad F (eds) The pharmacological basis of therapeutics, 7th edn. Macmillan, New York, pp 387–445
Baldessarini RJ, Cohen BM, Teicher MH (1988) Significance of neuroleptic dose and plasma level in the pharmacological treatment of psychosis. Arch Gen Psychiatry 45: 79–91
Bannon MJ, Bunney EB, Zigun JR, Skirboll LR, Roth RH (1980) Presynaptic dopamine receptors: insensitivity to kainic acid and the development of supersensitivity following chronic haloperidol. Naunyn-Schmiedebergs Arch Pharmacol 312: 161–164
Bannon MJ, Michaud RL, Roth RH (1981) Mesocortical dopamine neurons: lack of autoreceptors modulating dopamine synthesis. Mol Pharmacol 19: 270–275
Bannon MJ, Reinhard JF Jr, Bunney EB, Roth RH (1982) Unique response to antipsychotic drugs is due to the absence of terminal autoreceptors in mesocortical dopamine neurons. Nature (London) 296: 444–446
Barbieto L, Chéramy A, Godeheu G, Desce JM, Glowinski J (1990) Glutamate receptors of the quisqualate-kainate subtype are involved in the presynaptic regulation of the dopamine release in the cat caudate nucleus in vivo. Eur J Neurosci 2: 304–311
Benes FM, Bird ED (1987) An analysis of the arrangement of neurons in the cingulate cortex of schizophrenic patients. Arch Gen Psychiatry 44: 608–616
Bernheimer H, Birkmayer W, Hornykiewicz O, Jellinger K, Seitelberger F (1973) Brain dopamine and the syndromes of Parkinson and Huntington: clinical, morphological and neurochemical correlations. J Neurol Sci 20: 415–455
Biggio G, Casu M, Klimek V, Gessa GL (1980) Dopamine synthesis: tolerance to haloperidol and supersensitivity to apomorphine depend on pre-synaptic receptors. In: Cattabeni F, Racagni G, Spano PF, Costa E (eds) Advances in biochemical psychopharmacology, vol 24. Raven Press, New York, pp 18–22
Blaha CD, Lane RF (1987) Chronic treatment with classical and atypical antipsychotic drugs differentially decreases dopamine release in striatum and nucleus accumbens in vivo. Neurosci Lett 78: 199–204
Bogerts B (1985) Evidence for structural changes in the limbic system in schizophrenia. In: Shagass C, Josiassen RC, Bridger WH, Weiss KJ, Stoff D, Simpson GM (eds) Biological psychiatry. Elsevier, New York, pp 1015–1020
Bogerts B, Meertz E, Schonfeld-Bausch R (1985) Basal ganglia and limbic system pathology in schizophrenia. A morphometric study of brain volume and shrinkage. Arch Gen Psychiatry 42: 784–791
Bowers MB (1973) 5-HIAA and HVA following probenecid in acute psychotic patients treated with phenothiazines. Psychopharmacologia 28: 309–312

Bowers MB, Swigar ME, Jatlow PI Goicoechea N (1984) Plasma catecholamine metabolites and early response to haloperidol. J Clin Psychiatry 45: 248–251
Bowers MB, Swigar ME, Jatlow PI Hoffman FJ (1989) Plasma catecholamine metabolites and treatment response at neuroleptic steady state. Biol Psychiatry 25: 734–738
Brown WA, Herz LR (1989) Response to neuroleptic drugs as a device for classifying schizophrenia. Schizophr Bull 15: 123–129
Buchsbaum MS, Wu JC, DeLisi LE, Holcomb HH, Hazlett E, Cooper-Langston K, Kessler R (1987) Positron emission tomography studies of basal ganglia and somatosensory cortex neuroleptic drug effects: differences between normal controls and schizophrenic patients. Biol Psychiatry 22: 479–494
Bunney BS, Aghajanian GK (1978) Mesolimbic and mesocortical dopaminergic systems: physiology and pharmacology. In: Lipton MA, DiMascio A (eds) Psychopharmacology: a generation of progress. Raven Press, New York, pp 159–169
Bunney BS, Grace AA (1978) Acute and chronic haloperidol treatment: comparison of effects on nigral dopaminergic cell activity. Life Sci 23: 1715–1728
Bunney BS, Grace AA (1979) Effects of chronic haloperidol treatment on nigral dopaminergic cell activity. In: Usdin E, Kopin IJ, Barchas J (eds) Catecholamines: basic and clinical frontiers. Pergamon Press, New York, pp 666–668
Bunney BS, Walters JR, Roth RH, Aghajanian GK (1973) Dopaminergic neurons: effect of antipsychotic drugs and amphetamine on single cell activity. J Pharmacol Exp Ther 185: 560–571
Burt DR, Creese I, Snyder SH (1977) Antischizophrenic drugs: chronic treatment elevates dopamine receptor binding in brain. Science 196: 326–328
Bustos G, Roth RH (1972) Effect of γ-hydroxybutyrate on the release of monoamines from the rat striatum. Biochem Pharmacol 21: 2649–2652
Carlsson A, Lindqvist M (1963) Effect of chlorpromazine or haloperidol on formation of 3-methoxytyramine and normetanephrine in mouse brain. Acta Pharmacol Toxicol 20: 140–144
Carlsson A, Lindqvist M, Magnusson T, Waldeck B (1958) On the presence of 3-hydroxytyramine in brain. Science 127: 471
Carlsson A, Fuxe K, Hamberger B, Lindqvist M (1966) Biochemical and histochemical studies on the effects of imipramine-like drugs and (+)-amphetamine on central and peripheral catecholamine neurons. Acta Physiol Scand 67: 481–497
Charpentier P (1950) 10-(Dialkylaminoalkyl)-phenothiazines. US Patent 2, 519, 886
Chen J, Paredes W, Gardner EL (1991) Chronic treatment with clozapine selectively decreases basal dopamine release in nucleus accumbens but not in caudate-putamen as measured by in vivo brain microdialysis: further evidence for depolarization block. Neurosci Lett 122: 127–131
Chéramy A, Romo R, Godeheu G, Baruch P, Glowinski J (1986) In vivo presynaptic control of dopamine release in the cat caudate nucleus. II. Facilitatory or inhibitory influence of L-glutamate. Neuroscience 19: 1081–1090
Chesselet MF, Chéramy A, Romo R, Desban M, Glowinski J (1983) GABA in the thalamic motor nuclei modulates dopamine release from two dopaminergic nigro-striatal pathways in the cat. Exp Brain Res 51: 275–282
Chiodo LA, Bunney BS (1983) Typical and atypical neuroleptics: differential effects of chronic administration on the activity of A9 and A10 midbrain dopaminergic neurons. J Neurosci 3: 1607–1619
Chiodo LA, Bunney BS (1985) Possible mechanisms by which repeated clozapine administration differentially affects the activity of two subpopulations of midbrain dopamine neurons. J Neurosci 5: 2539–2544
Church WH, Justice JB, Neill DB (1987) Detecting behaviorally relevant changes in extracellular dopamine with microdialysis. Brain Res 421: 397–399

Clow DW, Jhamandas K (1989) Characterization of L-glutamate action on the release of endogenous dopamine from the rat caudate-putamen. J Pharmacol Exp Ther 248: 722–728

Connell PH (1958) Amphetamine psychosis. Chapman & Hall, London

Conrad AJ Abebe T, Austin R, Forsythe S, Scheibel AB (1991) Hippocampal pyramidal cell disarray in schizophrenia as a bilateral phenomenon. Arch Gen Psychiatry 48: 413–417

Cook L, Catania AC (1964) Effects of drugs on avoidance and escape behavior. Fed Proc 23: 818–835

Coyle S, Napier TC, Breese GR (1985) Ontogeny of tolerance to haloperidol: behavioral and biochemical measures. Dev Brain Res 23: 27–38

Creese I, Burt DR, Snyder SH (1976) Dopamine receptor binding predicts clinical and pharmacological potencies of antischizophrenic drugs. Science 192: 481–483

Curtis DR, Phillis JW, Watkins JC (1960) The chemical excitation of spinal neurones by certain acidic amino acids. J Physiol (London) 150: 656–682

Davidson AB, Weidley E (1976) Differential effects of neuroleptic and other psychotropic agents on acquisition of avoidance in rats. Life Sci 18: 1279–1284

Davila R, Manero E, Zumarraga MA, Andia I, Schweitzer JW, Friedhoff AJ (1988a) Plasma homovanillic acid as a predictor of response to neuroleptics. Arch Gen Psychiatry 45: 564–567

Davila R, Zumarraga M, Friedhoff AJ, Miller JC (1988b) Characteristics of the adaptive aspects of the dopaminergic system. Psychopharmacol Bull 24: 338–340

Davis JM (1985) Antipsychotic drugs. In: Kaplan HI, Sadock BJ (eds) Comprehensive textbook of psychiatry, vol 2, 4th edn. Williams & Wilkins, Baltimore MD, pp 1481–1513

Deakin JFW, Slater P, Simpson MDC, Gilchrist AC, Skan W, Royston MC, Reynolds GP, Cross AJ (1989) Frontal-cortical and left temporal glutamatergic dysfunction in schizophrenia. J Neurochem 52: 1781–1786

De Keyser J, Claeys A, De Backer J-P, Ebinger G, Roels F, Vanguelin G (1988) Autoradiographic localization of the D_1 and D_2 dopamine receptors in human brain. Neurosci Lett 91: 142–147

del Rio J, Fuentes JA (1969) Further studies on the antagonism of stereotyped behavior induced by amphetamine. Eur J Pharmacol 8: 73–78

Delay J, Deniker P, Harl JM (1952) Utilization en thérapeutique psychiatrique d'une phénothiazine d'action centrale élective (4560 RP). Ann Méd Psychol 110: 112–117

Delay J, Deniker P, Tardieu Y, Lemperiere T (1955) Premiers essais de la réserpine. Presse Méd 63: 663–665

Deniker P (1960) Experimental neurological syndromes and the new drug therapies in psychiatry. Compr Psychiatry 1: 92–102

Deutch AY, Roth RH (1988) Alterations in dopamine synthesis induced by chronic neuroleptic administration: a possible biochemical correlate of depolarization inactivation. Soc Neurosci Abstr 14: 27

Di Chiara G, Porceddu ML, Spano PF, Gessa GL (1977) Haloperidol increases and apomorphine decreases striatal dopamine metabolism after destruction of striatal dopamine-sensitive adenylate-cyclase by kainic acid. Brain Res 130: 374–382

Doherty MD, Gratton A (1991) Behavioral evidence of depolarization block of mesencephalic dopamine neurons by acute haloperidol in partially 6-hydroxydopamine lesioned rats. Behav Neurosci 105: 579–587

Doucet G, Descarries I, Garcia S (1986) Quantification of the dopamine innervation in adult rat neostriatum. Neuroscience 19: 427–445

Ewing AG, Wightman RM (1984) Monitoring the stimulated release of dopamine with in vivo voltammetry. II. Clearance of released dopamine from extracellular fluid. J Neurochem 43: 570–577

Fann WE, Lake CR (1976) Amantadine versus trihexylphenidyl in the treatment of neuroleptic-induced parkinsonism. Am J Psychiatry 8: 940–943

Fardé L, Wiesel FA, Nordstrom A-L, Sedvall G (1989) D1-and D2-dopamine receptor occupancy during treatment with conventional and atypical neuroleptics. Psychopharmacology 99: S28–S31

Farnebo LO, Hamberger B (1971) Drug-induced changes in the release of ^{3}H-monoamines from field-stimulated rat brain slices. Acta Physiol Scand [Suppl] 371: 35–44

Fibiger HC, LePiane FG, Jakubovic A, Phillips AG (1987) The role of dopamine in intracranial self-stimulation of the ventral tegmental area. J Neurosci 7: 3888–3896

Finlay JM, Jakubovic A, Fu DS, Fibiger HC (1987) Tolerance to haloperidol-induced increases in dopamine and metabolites: fact or artifact? Eur J Pharmacol 137: 117–121

Floran B, Silva I, Nava C, Aceves J (1988) Presynaptic modulation of the release of GABA by $GABA_A$ receptors in pars compacta and by $GABA_B$ receptors in pars reticulata of the rat substantia nigra. Eur J Pharmacol 150: 277–286

Flugel P (1956) Therapeutique par medication neuroleptiques obtenu en realizant systematique des etats Parkinsoniformes. Encéphale 45: 1090–1092

Frey JM, Ticku MK, Huffman RD (1987) GABAergic supersensitivity within the pars reticulata of the rat substantia nigra following chronic haloperidol administration. Brain Res 425: 73–84

Frey JM, Ticku MK, Bell RD, Huffman RD (1989) Chronic haloperidol administration increases GABA binding and enhances neuronal responsiveness to iontophoresed GABA in rat globus pallidus. Brain Res 491: 57–67

Friedhoff AJ, Miller JC (1983) Clinical implications of receptor sensitivity modification. Annu Rev Neurosci 6: 121–148

Friedhoff AJ, Bonnet K, Rosengarten H (1977) Reversal of two manifestations of dopamine receptor supersensitivity by administration of L-dopa. Res Commun Chem Pathol Pharmacol 116: 411–423

Gale K (1980) Chronic blockade of dopamine receptors by antischizophrenic drugs enhances GABA binding in substantia nigra. Nature (London) 283: 569–570

Garcia-Muñoz M, Nicolaou NM, Tulloch IF, Wright AK, Arbuthnot GW (1977) Feedback loop or output pathway in striatonigral fibers? Nature (London) 265: 363–365

Geffen LB, Jessell TM, Cuello AC, Iversen LL (1976) Release of dopamine from dendrites in substantia nigra. Nature 260: 258–260

Gerlach J, Luhdorf K (1975) The effect of L-DOPA on young patients with simple schizophrenia treated with neuroleptic drugs: a double-blind crossover trial with madopar and placebo. Psychopharmacology 44: 105–110

Giralt MT, Bonanno G, Raiteri M (1990) GABA terminal autoreceptors in the pars compacta and in the pars reticulata of the rat substantia nigra are $GABA_B$. Eur J Pharmacol 175: 137–144

Glazer WM, Bowers JB Jr, Charney DS, Heninger GR (1989) The effect of neuroleptic discontinuation on psychopathology, involuntary movements, and biochemical measures in patients with persistent tardive dyskinesia. Biol Psychiatry 26: 224–233

Glowinski J, Axelrod J, Iversen LL (1966) Regional studies of catecholamines in the rat brain. IV. Effects of drugs on the disposition and metabolism of ^{3}H-norepinephrine and ^{3}H-dopamine. J Pharmacol Exp Ther 153: 30–41

Glowinski J, Chéramy A, Romo R, Barbeito L (1988) Presynaptic regulation of dopaminergic transmission in the striatum. Cell Mol Neurobiol 8: 7–17

Goldstein JM, Litwin LC (1988) Spontaneous activity of A9 and A10 dopamine neurons after acute and chronic administration of the selective dopamine D-1 receptor antagonist SCH23390. Eur J Pharmacol 155: 175–180

Gonon FG (1988) Nonlinear relationship between impulse flow and dopamine released by rat midbrain dopaminergic neurons as studied by in vivo electrochemistry. Neuroscience 24: 19–28

Grace AA (1991) Phasic versus tonic dopamine release and the modulation of dopamine system responsivity: a hypothesis for the etiology of schizophrenia. Neuroscience 41: 1–24

Grace AA, Bunney BS (1979) Paradoxical GABA excitation of nigral dopaminergic cells: indirect mediation through reticulata inhibitory neurons. Eur J Pharmacol 59: 211–218

Grace AA, Bunney BS (1980a) Nigral dopamine neurons: intracellular recording and identification with L-DOPA injection and histofluorescence. Science 210: 654–656

Grace AA, Bunney BS (1980b) Effects of baclofen on nigral dopaminergic cell activity following acute and chronic haloperidol treatment. Brain Res Bull 5 [Suppl] 2: 537–543

Grace AA, Bunney BS (1983a) Intracellular and extracellular electrophysiology of nigral dopamine neurons. I. Identification and characterization. Neuroscience 10: 301–315

Grace AA, Bunney BS (1983b) Intracellular and extracellular electrophysiology of nigral dopaminergic neurons. II. Action potential generating mechanisms and morphological correlates. Neuroscience 10: 317–331

Grace AA, Bunney BS (1983c) Intracellular and extracellular electrophysiology of nigral dopaminergic neurons. III. Evidence for electrotonic coupling. Neuroscience 10: 333–348

Grace AA, Bunney BS (1984a) The control of firing pattern in nigral dopamine neurons: single spike firing. J Neurosci 4: 2866–2876

Grace AA, Bunney BS (1984b) The control of firing pattern in nigral dopamine neurons: burst firing. J Neurosci 4: 2877–2890

Grace AA, Bunney BS (1985a) Opposing effects of striatonigral feedback pathways on midbrain dopamine cell activity. Brain Res 333: 271–284

Grace AA, Bunney BS (1985b) Low doses of apomorphine elicit two opposing influences on dopamine cell electrophysiology. Brain Res 333: 285–298

Grace AA, Bunney BS (1986) Induction of depolarization block in midbrain dopamine neurons by repeated administration of haloperidol: analysis using in vivo intracellular recording. J Pharmacol Exp Ther 238: 1092–1100

Gratton A, Hoffer BJ, Gerhardt GA (1988) Effects of electrical stimulation of brain reward sites on release of dopamine in rat: an in vivo electrochemical study. Brain Res Bull 21: 319–324

Groves PM, Wilson CJ, Young SJ, Rebec GV (1975) Self-inhibition of dopaminergic neurons. Science 190: 522–529

Guidotti A, Gale K, Toffano G, Vargas FM (1978) Tolerance to tyrosine hydroxylase activation in n. accumbens and c. striatum after repeated injections of "classical" and "atypical" antischizophrenic drugs. Life Sci 23: 501–506

Hechler V, Gobaille S, Bourguignon J-J, Maitre M (1991) Extracellular events induced by γ-hydroxybutyrate in striatum: a microdialysis study. J Neurochem 56: 938–944

Heffner TG, Zigmond MJ, Stricker EM (1977) Effects of dopaminergic agonists and antagonists on feeding in intact and 6-hydroxydopamine-treated rats. J Pharmacol Exp Ther 201: 386–399

Herrling PL (1985) Pharmacology of the corticocaudate excitatory postsynaptic potential in the cat: evidence for its mediation by quisqualate or kainate receptors. Neuroscience 14: 417–426

Hollerman JR, Grace AA (1989a) Acute haloperidol administration induces depolarization block of nigral dopamine neurons in rats after partial dopamine lesions. Neurosci Lett 96: 82–88

Hollerman JR, Grace AA (1989b) Altered responsiveness to haloperidol following partial dopamine depletion in rats: behavioral and electrophysiological aspects. Soc Neurosci Abstr 15: 1002
Hollerman JR, Abercrombie ED, Grace AA (1992) Electrophysiological, biochemical, and behavioral studies of acute haloperidol-induced depolarization block of nigral dopamine neurons. Neuroscience 47: 589–601
Ichikawa J, Meltzer HY (1991) Differential effects of repeated treatment with haloperidol and clozapine on dopamine release and metabolism in the striatum and the nucleus accumbens. J Pharmacol Exp Ther 256: 348–357
Imperato I, DiChiara G (1985) Dopamine release and metabolism in awake rats after systemic neuroleptics as studied by trans-striatal dialysis. J Neurosci 5: 297–306
Imperato A, Honoré T, Jensen LH (1990) Dopamine release in the nucleus caudatus and accumbens is under glutamatergic control through non-NMDA receptors: a study in freely-moving rats. Brain Res 530: 223–228
Ingvar DH, Franzen G (1974) Abnormalities of cerebral blood flow distribution in patients with chronic schizophrenia. Acta Psychiatr Scand 50: 425–462
Invernizzi R, Morali F, Pozzi L, Samanin R (1990) Effects of acute and chronic clozapine on dopamine release and metabolism in the striatum and nucleus accumbens of conscious rats. Br J Pharmacol 100: 744–748
Iversen LL, Rogawski MA, Miller RJ (1976) Comparison of the effects of neuroleptic drugs on pre- and postsynaptic dopaminergic mechanisms in the rat striatum. Mol Pharmacol 12: 251–262
Iwatsubo K, Clouet D (1977) Effects of morphine and haloperidol on the electrical activity of rat nigrostriatal neurons. J Pharmacol Exp Ther 202: 429–436
Janssen AJ, Niemegeers CJE, Schellekens KHL (1965) Is it possible to predict the clinical effects of neuroleptic drugs (major tranquilizers) from animal data? Drug Res (Arzneimittelforschung) I15: 104–177
Javitt DC (1987) Negative symptomatology and the PCP (phencyclidine) model of schizophrenia. Hillside J Clin Psychiat 9: 12–35
Jiang LH, Tsai M, Wang RY (1988) Chronic treatment with high doses of haloperidol fails to decrease the time course for the development of depolarization inactivation of midbrain dopamine neurons. Life Sci 43: 75–81
Johnstone EC, Crow TJ, Frith CD, Carney MWP, Price JS (1978) Mechanism of the antipsychotic effect in the treatment of acute schizophrenia. Lancet i: 848–851
Kabzinski AM, Szewczak MR, Cornfeldt ML, Fielding S (1987) Differential effects of dopamine agonists and antagonists on the spontaneous electrical activity of A9 and A10 dopamine neurons. Soc Neurosci Abstr 13: 908
Kamata K, Sugimoto A, Kameyama T (1986) Effect of chronic haloperidol on dopamine release following microinjection of GABA into the substantia nigra zona reticulata in the rat. Brain Res 380: 1–6
Kane J, Honigfeld G, Singer J, Meltzer H (1988) Clozapine for the treatment-resistant schizophrenic: a double-blind comparison versus chlorpromazine/benztropine. Arch Gen Psychiatry 45: 789–796
Kebabian JW, Calne R (1979) Multiple receptors for dopamine. Nature 277: 93–96
Keefe KA, Stricker EM, Zigmond MJ, Abercrombie ED (1990) Environmental stress increases extracellular dopamine in striatum of 6-hydroxydopamine-treated rats: in vivo microdialysis studies. Brain Res 527: 350–353
Kehr W, Carlsson A, Lindqvist M, Magnusson T, Atack C (1972) Evidence for a receptor-mediated feedback control of striatal tyrosine hydroxylase. J Pharm Pharmacol 24: 744–747
Keller RW, Duhr WG, Wightman RM, Zigmond MJ (1988) The effect of L-DOPA on in vivo dopamine release from nigrostriatal bundle neurons. Brain Res 447: 191–194

Kim JS, Kornhuber HH, Schmid-Burgk W, Holzmiller B (1980) Low cerebrospinal fluid glutamate in schizophrenic patients and a new hypothesis of schizophrenia. Neurosci Lett 20: 379–382

Klein DE, Davis JM (1969) Diagnosis and drug treatment of psychiatric disorders. Williams & Wilkins, Baltimore MD

Kondo Y, Iwatsubo K (1980) Diminished responses of nigral dopaminergic neurons to haloperidol and morphine following lesions in the striatum. Brain Res 181: 237–240

Korf J, Venema K (1985) Amino acids in rat striatal dialysates: methodological aspects and changes after electroconvulsive shock J Neurochem 45: 1341–1348

Korf J, Zielman M, Westerink BHC (1976) Dopamine release in substantia nigra. Nature 260: 257–258

Kornhuber J, Mack-Burkhardt F, Riederer P, Hebenstreit GF, Reynolds GP, Andrews HB, Beckmann H (1989) [^{3}H]MK-801 binding sites in postmortem brain regions of schizophrenic patients. J Neural Transm 77: 231–236

Kuhr WG, Wightman RM (1986) Real-time measurement of dopamine release in rat brain. Brain Res 381: 168–171

Kuhr WG, Ewing AG, Caudill WL, Wightman RM (1984) Monitoring the stimulated release of dopamine with in vivo voltammetry. I. Characterization of the response observed in the caudate nucleus of the rat. J Neurochem 43: 560–569

Kuhr WG, Ewing AG, Near JA, Wightman RM (1985) Amphetamine attenuates the stimulated release of dopamine in vivo. J Pharmacol Exp Ther 232: 388–394

Laborit H, Huguenard P, Alluaume R (1952) Un nouveau stabilisateur végétatif, le 4560 RP. Presse Méd 60: 206

Lane RF, Blaha CD (1987) Chronic haloperidol decreases dopamine release in striatum and nucleus accumbens in vivo: depolarization block as a possible mechanism of action. Brain Res Bull 18: 135–138

Lavy S, Melamed E, Penchas S (1978) Tardive dyskinesia associated with metoclopramide. Br Med J 1: 77–78

Leviel V, Gobert A, Guibert B (1990) The glutamate-mediated release of dopamine in the rat striatum: Further characterization of the dual excitatory-inhibitory function. Neuroscience 39: 305–312

Leysen JE (1982) Review on neuroleptic receptors; specificity and multiplicity of in vitro binding relates to pharmacological activity. In: Usdin E, Dahl S, Gram LF, Lingjaerde O (eds) Clinical pharmacology in psychiatry: neuroleptic and antidepressant research. Macmillan, Basingstoke Hants, pp 35–47

Lidow MS, Goldman-Rakic, PS, Rakic P, Innis RB (1989) Dopamine D_2 receptors in the cerebral cortex: distribution and pharmacological characterization with [^{3}H]raclopride. Proc Natl Acad Sci USA 86: 6412–6416

Lindstrom LH (1985) Low HVA and normal 5-HIAA CSF levels in drug-free schizophrenic patients compared to healthy volunteers: correlations to symptomatology and family history. Psychiatry Res 14: 265–273

Lloyd KG, Davidson L, Hornykiewicz O (1975) The neurochemistry of Parkinson's disease: effect of L-dopa therapy. J Pharmacol Exp Ther 195: 453–464

Maas JW (1979) Neurotransmitters and depression. Too much, too tittle, or too unstable? Trends Neurosci 2: 306–308

Manos N, Gziouzepas J, Logothetis J (1981) The need for continuous use of antiparkinsonian medication with chronic schizophrenic patients receiving long-term neuroleptic therapy. Am J Psychiatry 138: 184–188

Marsden CD, Tarsy D, Baldessarini RJ (1975) Spontaneous and drug-induced movement disorders in psychotic patients. In: Benson DF, Blumer D (eds) Psychiatric aspects of neurologic disease. Grune & Stratton, New York, pp 219–266

Matthysse S (1973) Antipsychotic drug actions: a clue to the neuropathology of schizophrenia? Fed Proc 32: 200–205

May PRA, Tuma AH, Yale C, Potepam, P, Dixon WJ (1976) Schizophrenia: a follow-up study of results of treatment, hospital stay over two to five years. Arch Gen Psychiatry 33: 481–506

May LJ, Kuhr WG, Wightman RM (1988) Differentiation of dopamine overflow and uptake processes in the extracellular fluid of the rat caudate nucleus with fast-scan in vivo voltammetry. J Neurochem 51: 1060–1069

McGeorge AJ, Faull RLM (1989) The organization of the projection from the cerebral cortex to the striatum in the rat. Neuroscience 29: 503–537

Meltzer HY (1984) Dopamine and negative symptoms in schizophrenia: critique of the type I-II hypotheses. In: Alpert M (ed) Controversies in schizophrenia: changes and consistencies. Guilford Press, New York, pp 110–144

Meltzer HY (1989) Clinical studies on the mechanism of action of clozapine: the dopamine-serotonin hypothesis of schizophrenia. Psychopharmacology 99 [Suppl]: S18–S32

Meltzer HY (1991) The mechanism of action of novel antipsychotic drugs. Schizophr Bull 17: 263–287

Meltzer HY, Sommers AA, Luchins DJ (1986) The effect of neuroleptics and other psychotropic drugs on negative symptoms in schizophrenia. J Clin Pharmacol 6: 329–338

Meltzer LT, Christoffersen CL, Heffner TG, Freeman AS, Chiodo LA (1989) CI-943, a potential antipsychotic agent. III. Evaluation of effects on dopamine neuronal activity. J Pharmacol Exp Ther 251: 123–130

Mercuri NB, Calabresi P, Bernardi G (1990) Responses of rat substantia nigra compacta neurones to L-DOPA. Br J Pharmacol 100: 257–260

Mereu G, Casu M, Gessa GL (1983) (-)-Sulpiride activates the firing rate and tyrosine hydroxylase activity of dopaminergic neurons in unanesthetized rats. Brain Res 264: 105–110

Mishra RK, Marshall AM, Varmuza SL (1980) Supersensitivity in rat caudate nucleus: effects of 6-hydroxydopamine on the time course of dopamine receptor and cyclic AMP changes. Brain Res 200: 47–57

Moghaddam B, Gruen RJ (1991) Do endogenous excitatory amino acids influence striatal dopamine release? Brain Res 544: 329–330

Moghaddam B, Gruen RJ, Roth RH, Bunney BS, Adams RN (1990) Effect of L-glutamate on the release of striatal dopamine: in vivo dialysis and electrochemical studies. Brain Res 518: 55–60

Munkvad I, Pakkenberg H, Randrup A (1968) Aminergic systems in basal ganglia associated with stereotyped hyperactive behavior and catalepsy. Behav Brain Evol 1: 89–100

Nakahara D, Ozaki N, Kapoor V, Nagatsu T (1989) The effect of uptake inhibition on dopamine release from the nucleus accumbens of rats during self- or forced stimulation of the medial forebrain bundle. A microdialysis study. Neurosci Lett 104: 136–140

Nieoullon A, Chéramy A, Glowinski J (1978) Release of dopamine evoked by electrical stimulation of the motor and visual areas of the cerebral cortex in both caudate nuclei and in the substantia nigra in the cat. Brain Res 145: 69–83

Nisenbaum ES, Stricker EM, Zigmond MJ, Berger TW (1986) Long-term effects of dopamine-depleting brain lesions on spontaneous activity of type II striatal neurons: relation to behavioral recovery. Brain Res 398: 221–230

Nisenbaum ES, Berger TW, Grace AA (1991) $GABA_B$ receptors reduce both glutamatergic excitation and GABAergic inhibition of striatal neurons through presynaptic mechanisms. Soc Neurosci Abstr 17: 850

Nishikawa T, Takashima M, Toru M (1983a) Increased [^{3}H]kainic acid binding in the prefrontal cortex in schizophrenia. Neurosci Lett 40: 245–250
Nishikawa R, Tanaka M, Koga I, Uchida Y (1983b) Combined treatment of tardive dyskinesia with clonidine and neuroleptics: a follow-up study of three cases for three years. Psychopharmacology 80: 374–375
Nowycky MC, Roth RH (1977) Presynaptic dopamine receptors. Development of supersensitivity following treatment with fluphenazine decanoate. Naunyn-Schmiedebergs Arch Pharmacol 300: 247–254
Nybåck H, Berggen BM, Hindmarsh T, Sedvall G, Weisel FA (1983) Cerebroventricular size and cerebrospinal fluid monoamine metabolites in schizophrenic patients and healthy volunteers. Psychiatry Res 9: 301–308
O'Neill RD, Gonzalez-Mora JL, Boutelle MG, Ormonde DE, Lowry JP, Duff A, Fumero B, Fillenz M, Mas M (1991) Anomalously high concentrations of brain extracellular uric acid detected with chronically implanted probes — Implications for in vivo sampling techniques. J Neurochem 57: 22–29
Pan HS, Penney JB, Young AB (1985) γ-Aminobutyric acid and benzodiazepine receptor changes induced by unilateral 6-hydroxydopamine lesions of the medial forebrain bundle. J Neurochem 45: 1396–1404
Pani L, Gessa GL, Carboni S, Portas CM, Rossetti ZL (1990) Brain dialysis and dopamine: does the extracellular concentration of dopamine reflect synaptic release? Eur J Pharmacol 180: 85–90
Peroutka SJ, Snyder SH (1980) Relationship of neuroleptic drug effects at brain dopamine, serotonin, α-adrenergic, and histamine receptors to clinical potency. Am J Psychiatry 137: 1518–1522
Perry TL (1982) Normal cerebrospinal fluid and brain glutamate levels in schizophrenia do not support the hypothesis of glutamatergic neuronal dysfunction. Neurosci Lett 28: 81–85
Pettegrew JW, Keshavan M, Panchalingam K (1989) ^{31}P NMR studies in schizophrenia. Biol Psychiatry 25, 15A
Pettegrew JW, Keshavan MS, Panchalingam K, Strychor S, Kaplan DB, Tretta M, Allen M (1991) Alterations in brain high-energy phosphate and membrane phospholipid metabolism in first-episode, drug-naive schizophrenics. Arch Gen Psychiatry 48: 563–568
Pickar D, Labarca R, Linnoila M, Roy A, Hommer D, Everett D, Paul SM (1984) Neuroleptic-induced decrease in plasma HVA and antipsychotic activity in schizophrenic patients. Science 225: 954–957
Pickar D, Labarca R, Doran AR, Wolkowitz OM, Roy A, Breier A, Linnoila M, Paul SM (1986) Longitudinal measurement of plasma homovanillic acid levels in schizophrenic patients. Arch Gen Psychiatry 43: 669–676
Post RM, Fink E, Carpenter WT, Goodwin FK (1975) Cerebrospinal fluid amine metabolites in acute schizophrenia. Arch Gen Psychiatry 32: 1063–1069
Pucak ML, Grace AA (1991a) Partial dopamine depletions result in an enhanced sensitivity of residual dopamine neurons to apomorphine. Synapse 9: 144–155
Pucak ML, Grace AA (1991b) Blockade of somatodendritic autoreceptors on nigral dopamine neurons contributes to the firing rate-increasing effects of dopamine antagonists. Soc Neurosci Abstr 17: 1352
Randrup A, Munkvad I (1965) Special antagonism of amphetamine-induced abnormal behaviour. Psychopharmacologia 7: 416–422
Reimann W, Zumstein A, Jackisch R, Starke K, Hertting G (1979) Effect of extracellular dopamine on the release of dopamine in the rabbit caudate nucleus: evidence for a dopaminergic feed-back inhibition. Naunyn-Schmiedebergs Arch Pharmacol 306: 53–60

Reimann W, Zumstein A, Starke K (1982) γ-Aminobutyric acid can both inhibit and facilitate dopamine release in the caudate nucleus of the rabbit. J Neurochem 39: 961–969

Reivich M, Kuhl D, Wolf A, Greenberg J, Phelps M, Ido T, Casella V, Fowler J, Hoffman E, Alavi A, Som P, Sokoloff L (1979) The (18F) fluorodeoxyglucose method for the measurement of local cerebral glucose utilization in man. Circ Res 44: 127–137

Richfield EK, Penney JB, Young AB (1989) Anatomical and affinity state comparisons between dopamine D_1 and D_2 receptors in the rat central nervous system. Neuroscience 30: 767–777

Roberts GW, Bruton CJ (1990) Notes from the graveyard: neuropathology and schizophrenia. Neuropathol Appl Neurobiol 16: 3–16

Robinson TE, Whishaw IQ (1988) Normalization of extracellular dopamine in striatum following recovery from a partial unilateral 6-OHDA lesion of the substantia nigra: a microdialysis study in freely moving rats. Brain Res 450: 209–224

Romo R, Chéramy A, Godeheu G, Glowinski J (1986) In vivo presynaptic control of dopamine release in the cat caudate nucleus. I. Opposite changes in neuronal activity and release evoked from thalamic motor nuclei. Neuroscience 19: 1067–1079

Rompré P-P, Wise RA (1989) Behavioral evidence for midbrain dopamine neuron depolarization block. Brain Res 477: 152–156

Sah P, Hestrin S, Nicoll RA (1989) Tonic activation of NMDA receptors by ambient glutamate enhances excitability of neurons. Science 246: 815–818

Sands SB, Barish ME (1989) A quantitative description of excitatory amino acid neurotransmitter responses on cultured embryonic Xenopus spinal neurons. Brain Res 502: 375–386

Scatton B, Worms P (1978) Subsensitivity of striatal and mesolimbic dopamine target cells after repeated treatment with apomorphine dipivaloyl ester. Naunyn-Schmiedebergs Arch Pharmacol 303: 271–278

Sedvall G, Farde L, Persson A, Weisel FA (1986) Imaging of neurotransmitter receptors in the living human brain. Arch Gen Psychiatry 43: 995–1006

Seeman P, Chau-Wong M, Tedesco J, Wong K (1975) Brain receptors for antipsychotic drugs and dopamine: direct binding assays. Proc Natl Acad Sci USA 72: 4376–4380

Seeman P, Lee T, Chau-Wong M, Wong K (1976) Antipsychotic drug doses and neuroleptic/dopamine receptors. Nature (London) 261: 717–719

Seeman P, Bzowej NJ, Guan HC, Bergeron C, Reynolds GP, Bird ED, Riederer P, Jellinger K, Tourtelotte WW (1987) Human brain D_1 and D_2 dopamine receptors in schizophrenia, Alzheimer's, Parkinson's and Huntington's diseases. Neuropsychopharmacology 1: 5–15

Serrano A, D'Angio M, Scatton B (1989) NMDA antagonists block restraint-induced increase in extracellular DOPAC in rat nucleus accumbens. Eur J Pharmacol 162: 157–166

Shalaby I, Spear LP (1980) Psychopharmacological effects of low and high doses of apomorphine during ontogeny. Eur J Pharmacol 67: 451–459

Sharp T, Zetterstrom T, Ungerstedt U (1986) An in vivo study of dopamine release and metabolism in rat brain regions using intracerebral dialysis. J Neurochem 47: 113–122

Shelton RC, Karson CN, Doran AR, Pickar D, Bigelow LB, Weinberger DR (1988) Cerebral structural pathology in schizophrenia: evidence for a selective prefrontal cortical defect. Am J Psychiatry 145: 154–163

Sherman AD, Davidson AT, Baruah S, Hegwood TS, Waziri R (1991) Evidence of glutamatergic deficiency in schizophrenia. Neurosci Lett 121: 77–80

Siever LJ, Davis KL (1985) Overview: toward a dysregulation hypothesis of depression. Am J Psychiatry 142: 1017–1031

Skarsfeldt T (1988) Differential effects after repeated treatment with halopoeridol, clozapine, thioridazine, and tefludazine on SNC and VTA dopamine neurons in rats. Life Sci 42: 1037–1044

Skirboll LR, Grace AA, Bunney BS (1979) Dopamine auto- and postsynaptic receptors: electrophysiological evidence for differential sensitivity to dopamine agonists. Science 206: 80–82

Sneed OC, Bearden LJ (1980) Naloxone overcomes the dopaminergic, EEG, and behavioral effects of γ-hydroxybutyrate. Neurology 30: 832–838

Snell LD, Johnson KM (1985) Antagonism of N-methyl-D-aspartate-induced transmitter release in the rat striatum by phencyclidine-like drugs and its relationship to turning behavior. J Pharmacol Exp Ther 235: 50–57

Snyder SH (1972) Catecholamines in the brain are mediators of amphetamine psychosis. Arch Gen Psychiatry 27: 169–179

Snyder SH (1973) Amphetamine psychosis: a model schizophrenia mediated by catecholamines, Am J Psychiatry 130: 61–67

Snyder SH (1980) Biological aspects of mental disorders, chapter 5. Schizophrenia: etiology and treatment. Oxford University Press, New York, pp 58–68

Snyder AM, Stricker EM, Zigmond MJ (1985) Stress-induced neurological impairments in an animal model of Parkinsonism. Ann Neurol 18; 544–551

Sokoloff P, Giros B, Martres M-P, Bouthenet M-L, Schwartz J-C (1990) Molecular cloning and characterization of a novel dopamine receptor (D_3) as a target for neuroleptics. Nature (London) 347: 146–151

Sorensen SM, Humphreys TM, Palfreyman MG (1989) Effect of acute and chronic MDL 73147EF, a 5-HT_3 receptor antagonist, on A9 and A10 dopamine neurons. Eur J Pharmacol 163: 115–118

Spencer H (1976) Antagonism of cortical excitation of striatal neurones by glutamic acid diethylester: Evidence for glutamic acid as an excitatory transmitter in the rat striatum. Brain Res 102: 91–101

Stevens JR (1973) An anatomy of schizophrenia? Arch Gen Psychiatry 29: 177–189

Stevens JR (1991) Psychosis and the temporal lobe. In: Smith D, Treiman D, Trimble M (eds) Advances in neurology, vol 55. Raven Press, New York, pp 79–96

Stille G, Hippius H (1971) Kritische Stellungsnahme zum Begriff der Neuroleptika (anhand von pharmakologischen und klinischen Befunden mit Clozapin). Pharmacopsychiatry 4: 182–191

Strecker RE, Jacobs BL (1985) Substantia nigra dopaminergic unit activity in behaving cats: effect of arousal on spontaneous discharge and sensory evoked activity. Brain Res 361: 339–350

Tamminga CA, Schaffer MH, Smith RC, Davis JM (1978) Schizophrenic symptoms improve with apomorphine. Science 200: 567–568

Tamminga CA, Kaneda H, Buchanan R, Kirkpatrick B, Thaker GK, Yablinski MB, Holcomb HH (1991) The limbic system in schizophrenia. Pharmacologic and metabolic evidence. In: Tamminga CA, Schulz SC (eds) Advances in neuropsychiatry and psychopharmacology, vol 1. Schizophrenia research. CA Raven Press, New York, pp 99–109

Taylor KM, Snyder SH (1971) Differential effects of d- and l-amphetamine on behavior and on catecholamine disposition in dopamine and norepinephrine containing neurons of rat brain. Brain Res 28: 295–309

Torrey EF (1987) Prevalence studies in schizophrenia. Br J Psychiatry 150: 598–608

Tupin JP (1985) Focal neuroleptization: an approach to optimal dosing for initial and continuing therapy. J Clin Psychopharmacol 5 [Suppl] 3: 1521

Ungerstedt U (1971) Postsynaptic supersensitivity after 6-hydroxydopamine induced degeneration of the nigro-striatal dopamine system. Acta Physiol Scand 83 (S367): 95–122

Ungerstedt U, Herrera-Marschnitz M, Jungnelius U, Stahle L, Tossman U, Zetterström T (1982) Dopamine synaptic mechanisms reflected in studies combining behavioral recordings and brain dialysis. Adv Biosci 37: 219–231

Van Kammen DP, Mann LS, Sternberg DE, Scheinin M, Ninan PT, Marder SR, Van Kammen WB, Reider RO, Linnoila M (1983) Dopamine-β-hydroxylase activity and homovanillic acid in spinal fluid of schizophrenic with brain atrophy. Science 220: 974–977

Van Kammen DP, Van Kammen WB, Mann LS, Seppala T, Linnoila M (1986) Dopamine metabolism in the cerebrospinal fluid of drug-free schizophrenic patients with and without cortical atrophy. Arch Gen Psychiatry 43: 978–983

Van Rossum JM (1966) The significance of dopamine-receptor blockade for the actions of neuroleptic drugs. Arch Int Pharmacodyn Thèr 160: 492–494

Van Tol HHM, Bunzow JR, Guan H-C, Sunahara RK, Seeman P, Niznik HB, Civelli O (1991) Cloning of the gene for a human D_4 receptor with high affinity for the antipsychotic clozapine. Nature (London) 350: 610–614

Vickroy TW, Johnson KM (1983) Effects of phencyclidine on the release and synthesis of newly formed dopamine. Neuropharmacology 22: 839–842

Waldmeier PC (1991) The $GABA_B$ antagonist, CGP 35348, antagonizes the effects of baclofen, γ-butyrolactone and HA 966 on rat striatal dopamine synthesis. Naunyn-Schmiedebergs Arch Pharmacol 343: 173–178

Weick BG, Engber TM, Susel Z, Chase TN, Walters JR (1990) Responses of substantia nigra pars reticulata neurons to GABA and SKF 38393 in 6-hydroxydopamine-lesioned rats are differentially affected by continuous and intermittent levodopa administration. Brain Res 523: 16–22

Weinberger DR, Berman KF, Zec RF (1986) Physiological dysfunction of dorsolateral prefrontal cortex in schizophrenia. Arch Gen Psychiatry 43: 114–124

White FJ, Wang RY (1983a) Comparison of the effects of chronic haloperidol treatment on A9 and A10 dopamine neurons in the rat. Life Sci 32: 983–993

White FJ, Wang RY (1983b) Differential effects of classical and atypical antipsychotic drugs on A9 and A10 dopamine neurons. Science 221: 1054–1057

Williamson P, Drost D, Stanley J, Carr T, Morrison S, Merskey H (1991) Localized phosphorus 31 magnetic resonance spectroscopy in chronic schizophrenic patients and normal controls. Arch Gen Psychiatry 48: 578

Wood GD (1977) An adverse reaction to metoclopramide therapy. Br J Oral Surg 15: 278–280

Wuerthele SM, Freed WJ, Olson L, Morihisa J, Spoor L, Wyatt RJ, Hoffer BJ (1981) Effect of dopamine agonists and antagonists on the electrical activity of substantia nigra neurons transplanted into the lateral ventricle of the rat. Exp Brain Res 44: 1–10

Wyatt RJ (1991) Neuroleptics and the natural course of schizophrenia. Schizophr Bull 17: 325–351

Zetterström T, Sharp T, Marsden CA, Ungerstedt U (1983) In vivo measurement of dopamine and its metabolites by intracerebral dialysis: changes after d-amphetamine. J Neurochem 41: 1769–1773

Zetterström T, Sharp T, Ungerstedt U (1984) Effect of neuroleptic drugs on striatal dopamine release and metabolism in the awake rat studied by intracerebral dialysis. Eur J Pharmacol 106: 27–37

Zetterström T, Sharp T, Ungerstedt U (1985) Effect of neuroleptic drugs on striatal dopamine release and metabolism in the awake rat studied by intracerebral dialysis. Eur J Pharmacol 106: 27–37

Zetterström T, Sharp T, Collin AK, Ungerstedt U (1988) In vivo measurement of extracellular dopamine and DOPAC in rat striatum after various dopamine-releasing drugs: implications for the origin of extracellular DOPAC. Eur J Pharmacol 148: 327–334

Zigmond MJ, Stricker EM (1973) Recovery of feeding and drinking by rats after intraventricular 6-hydroxydopamine or lateral hypothalamic lesions. Science 182: 717–720

Zigmond MJ, Berger TW, Grace AA, Stricker EM (1989) Compensatory responses to nigrostriatal bundle injury: studies with 6-hydroxydopamine in an animal model of Parkinsonism. Mol Chem Neuropathol 10: 185–200

Zigmond MJ, Abercrombie ED, Berger TW, Grace AA, Stricker EM (1990) Compensations after lesions of central dopaminergic neurons: some clinical and basic implications. Trends Neurosci 13: 290–296

Author's address: Dr. A. A. Grace, Departments of Behavioral Neuroscience and Psychiatry, University of Pittsburgh, 458 Crawford Hall, Pittsburgh, PA 15260, U.S.A.

Prefrontal cortex

J Neural Transm (1992) [Suppl] 36: 135–162

NE/DA interactions in prefrontal cortex and their possible roles as neuromodulators in schizophrenia

J. P. Tassin

Chaire de Neuropharmacologie, INSERM U.114, Collège de France, Paris, France

Summary. The monoaminergic innervation of the rat prefrontal cortex arises from well-defined mesencephalic nuclei, with noradrenergic (NE) neurons located in the locus coeruleus, dopaminergic (DA) neurons located in the ventral tegmental area, and serotonergic (5-HT) neurons originating in the raphe nuclei. Specific destruction of the NE bundle was found to induce morphological (i.e., sprouting) as well as metabolic (i.e., changes in rate of DA utilization) modifications of mesocortical DA neurons, suggesting that these two catecholaminergic systems have functional interactions within the prefrontal cortex. This was substantiated by experiments showing that DA afferents modulate the sensitivity of cortical post-synaptic β-adrenergic receptors and that, reciprocally, NE neurons control the sensitivity of cortical D1 receptors. Behavioural and pharmacological data have further indicated that the stimulation of cortical alpha-1 adrenergic receptors inhibits cortical DA transmission at D1 receptors.

Secondly, we have attempted to analyze how such interactions between neuromodulatory systems may be related to the development of mental diseases such as schizophrenia. On the basis of studies in the literature describing the effects produced by the ingestion of hallucinogenic drugs or data collected regarding REM sleep, it is postulated that two modes of brain functioning exist: analogical and cognitive. Each mode is characterized by differences in the relative activities of NE, DA and 5-HT neurons. At birth, during REM sleep, and following the ingestion of hallucinogens, the mode of brain functioning is essentially analogical; in contrast, both analogic and cognitive modes are postulated to coexist in the awake state. Oscillations between these two modes are under the control of monoaminergic systems on which an increase in cortical DA release favours the cognitive processing mode, whereas intermittent activations of NE neurons would switch the brain into the analogical mode of processing. It is proposed that schizophrenic patients with "positive" symptoms suffer from an abnormal preponderance of the analogical mode while awake, whereas "negative" symptoms are due to the excessive presence of the cognitive mode. Although pure biological deficits cannot be excluded, these dysfunctions could be related to the absence of particular environmental variables early in the development of these patients. This condition is probably required to establish normal regulatory control of monoaminergic neuronal activity.

Introduction

The rat prefrontal cortex receives a dopaminergic (DA) innervation that arises primarily from cell bodies located in the rostro-medial part of the ventral tegmental area (VTA) (Fuxe et al., 1974). This cortical DA innervation is particularly dense in the deeper layers of the cortex (Berger et al., 1974; Björklund and Lindvall, 1978). This region of the cortex is also the locus of the D1 subtype of DA receptors (Von Hungen and Roberts, 1973; Bockaert et al., 1977). The topographical distribution of these D1 receptors in the anterior cerebral cortex matches closely that of the afferent DA terminals, and neurotoxic lesion studies have shown that these receptors are located postsynaptically on cortical cells in this region (Tassin et al., 1978a, 1982a).

On the other hand, noradrenergic (NE) fibers, which originate in the locus coeruleus, are distributed more extensively in the superficial layers of the cortex, including areas corresponding to the DA projection field (Swanson and Hartman, 1975; Morrison et al., 1981). Cortical neurons located in this region of overlap have been shown to be sensitive to microiontophoretic application of both DA and NE (Bunney and Aghajanian, 1976), suggesting that DA/NE interactions may occur in this anterior part of the cerebral cortex.

Many reports have provided evidence for interactions between NE and DA systems in the central nervous system (Antelman and Caggiula, 1977; Cools et al., 1978; Tassin et al., 1978b; Carter and Pycock, 1980; Van Kammen and Antelman, 1984). These interactions, which can be demonstrated by biochemical analysis, have been shown to exert significant pharmacological and behavioural effects. For example, Arnsten and Goldman-Rakic (1985) have shown that the cognitive deficits that occur in monkeys during aging, which are thought to be due to progressive loss of DA innervation, can be ameliorated by administering clonidine. Similarly, it has been shown that alpha-1 and alpha-2 adrenoreceptor antagonists can affect the behaviour induced by administering DA agonists (Dickinson et al., 1988).

For several years, the dorsal NE bundle was considered to be the only catecholaminergic pathway innervating the cerebral cortex. However, lesion experiments showed that cortical DA levels of rats depleted of NE were not decreased, suggesting the presence of a distinct DA input to the cortex (Thierry et al., 1973). Further experiments, also involving the destruction of the dorsal NE bundle, showed a decrease in DA utilization and an increase in the density of DA nerve terminals in the prefrontal cortices which are depleted of NE. These results again support an interaction between these two catecholaminergic systems (Tassin et al., 1978b).

The discovery of mesocortical DA neurons was of particular significance, since it provided new insight into the DA hypothesis of schizophrenia. In its original form, the DA hypothesis emphasized overactivity primarily in the DA system innervating subcortical limbic structures. Thus, it was thought that the antipsychotic effects of neuroleptics were linked primarily to their ability to block DA transmission within these subcortical structures, which included the anterior striatum, nucleus accumbens, and amygdala. Nonetheless, functional deficits reminiscent of those seen in schizophrenic patients, such as disruption of higher cognitive functions, attention deficits, increased and stereotypic motor behavior, and disturbance in affective response were observed in patients with lesions in the prefrontal cortex. Thus a dysfunction of the mesocortical DA system may better account for some of the characteristic symptoms of this disease. The observation that the destruction of ascending NE fibers modifies mesocortical DA transmission seemed, therefore, of particular interest, especially as part of the therapeutic efficacy of neuroleptics may be related to their action on NE transmission via their blockade of alpha-1 adrenergic receptors (Seeman and Grigoriadis, 1987).

In this review, I will summarize some of our studies which show the presence of multiple levels of interaction between NE and DA transmission in the prefrontal cortex. First, we will show that ascending NE neurons can modulate the activity and the development of mesocortical DA neurons. We will then describe experiments demonstrating interactions between postsynaptic receptors of both catecholaminergic systems. Indeed, it will be shown that mesocortical DA neurons modulate the sensitivity of the prefrontocortical β-adrenergic receptors and that, reciprocally, NE neurons regulate cortical DA receptors. More precisely, the presence of an intact NE innervation seems to be necessary to obtain denervation hypersensitivity of prefrontal D1 DA receptors. Finally, some experiments showing the behavioural consequences of these receptor heteroregulations will be presented. These studies suggest that the blockade of alpha-1 adrenergic receptors facilitates cortical D1 transmission and further emphasizes the possible role of the blockade of these NE receptors in the therapeutic efficacy of neuroleptics.

Neuroleptics, however, also interfere with serotonergic (5-HT) transmission via the blockade of 5-HT2 and 5-HT1A receptors (Seeman and Grigioradis, 1987; Wander et al., 1987). In addition, numerous studies have shown that 5-HT neurons can influence catecholaminergic transmission. Brunello et al. (1982), for example, have described the permissive role of 5-HT axons on the down-regulation of cortical β-adrenergic receptors following the chronic blockade of NE reuptake by an antidepressant. Similarly, stimulation of 5-HT receptors inhibits DA release in the striatum (Ennis et al., 1981). Obviously, a comprehensive study of neuromodulator interactions should not be limited to catecholamines, but should also consider the role of 5-HT. Since this workshop is devoted to schizophrenia and is directed at forming bridges between basic and clinical sciences, we will try, in the second part of this discussion, to extrapolate our data in this direction. Furthermore, by comparing the functional characteristics of NE, DA, and 5-HT transmission, we plan to show how a better knowledge of the interactions among these three neuromodulators may help to understand the mechanisms underlying this disorder.

NE/DA interactions in the prefrontal cortex

Collateral sprouting and reduced activity of the mesocortical DA neurons following the destruction of ascending NE fibers

Striking modifications in the properties of the mesocortical DA neurons projecting to the anteromedial cerebral cortex of the rat were seen after degeneration of the ascending NE pathways. Cortical DA levels were increased by up to 70% when measured 3 to 6 months following bilateral 6-OHDA lesions of NE neurons. Furthermore, when examined in individual animals, the alteration in DA levels correlated with the increase in cortical 3H-DA high affinity uptake (Tassin et al., 1978b). Histochemical analyses confirmed the presence of an increased density of DA nerve terminals in some cortical areas following the degeneration of NE neurons. The sprouting process involved both the thin (pregenual) and thick (supragenual) fibers originating from the VTA and the SN respectively (Fallon and Moore, 1978; Tassin et al., 1978b).

Interestingly, this increased cortical DA innervation was accompanied by a 50% decrease in the dihydroxyphenylacetic acid (DOPAC)/DA ratios in prefrontal cortices of animals devoid of NE terminals. These decreases in cortical DA utilization were constant across all groups of lesioned animals and were therefore not dependent on changes in cortical DA levels. Indeed, animals with almost no increase in cortical DA levels still exhibited 50% decreases in their cortical DOPAC/DA ratios (Tassin et al., 1978b).

It seems likely that the collateral sprouting of the cortical DA fibers was due to the disappearance of the cortical NE innervation. However, since our 6-OHDA lesions destroyed both the dorsal and the ventral NE pathways, the decrease in cortical DA utilization could be induced by NE/DA interactions at either of two levels: 1) at the nerve terminals in the prefrontal cortex, or 2) at the DA cell bodies within the VTA. To test this hypothesis, 6-OHDA lesions were made in the vicinity of the NE pathway connecting the locus coeruleus to the VTA and corresponding to the dorsal part of the decussatio of the pedunculus cerebellaris superior (P.C.S.) in the medial region of the tegmental radiation (Hervé et al., 1982). These lesions, which reduced NE levels by 64% in the VTA, only slightly affected NE levels in the prefrontal cortex (−19%). However, they induced a 38% decrease in the DOPAC/DA ratio in the prefrontal cortex. Since no change in DA utilization was observed in the nucleus accumbens of these lesioned animals, the NE/DA interaction in the VTA appeared to be limited to the mesocortical DA neurons. The results also suggest that interactions between ascending NE fibers and mesocortical DA neurons occur not only in the prefrontal cortex but also in the VTA (Hervé et al., 1982).

Involvement of mesocortical DA neurons in the regulation of β-adrenergic receptor sensitivity in rat prefrontal cortex

The contribution of DA afferents to the regulation of β-adrenergic receptor sensitivity, as defined by isoproterenol-stimulated adenylate cyclase activity in the rat prefrontal cortex was investigated by comparing the effects of lesioning only the NE fibers with those produced by lesioning both the DA and NE neurons. Bilateral 6-OHDA lesions of the VTA destroyed ascending DA and, to a variable extent, ascending NE fibers innervating the prefrontal cortex (Tassin et al., 1982b). Consistent with previous studies (Sporn et al., 1976), when the NE denervation was complete, a marked increase in the isoproterenol-stimulated adenylate cyclase activity (+78%) was found. However, when 6-OHDA injections into the VTA led to a partial destruction of cortical NE afferents (to 10–40% of control values), an isoproterenol-induced adenylate cyclase activity decreased by 30%. This effect contrasted with the β-adrenergic receptor hypersensitivity seen in rats which had received bilateral 6-OHDA lesions laterally into the P.C.S. causing partial NE denervations without affecting cortical DA levels. The hyposensitivity of β-adrenergic receptors obtained in rats with partial lesions of cortical NE fibers and devoid of cortical DA innervation, therefore, suggests that the mesocortical DA neurons may regulate β-adrenergic receptor density when the activity of ascending NE neurons is partially decreased (Hervé et al., 1990).

Influence of the cortical NE innervation on the regulation of D1 receptors in the prefrontal cortex

As mentioned previously, lesion experiments of mesocortical DA neurons have indicated that cortical D1 receptors are post-synaptic to the cortical DA afferents. Indeed, 6-OHDA injections into the VTA did not produce any change in the activity of the prefronto-cortical D1-linked adenylate cyclase, even though cortical DA levels were reduced by more than 98%. These results further indicated that, in the central nervous system, even the total destruction of a homogeneous population of afferent presynaptic fibers does not necessarily produce denervation hypersensitivity of the corresponding post-synaptic receptors (Tassin et al., 1982a). However, when the VTA lesions were performed electrolytically, the DA-sensitive adenylate cyclase activity (measured in tissue homogenates) and the density of D1 receptors (obtained by quantitative autoradiography) increased by 38% and 24%, respectively (Tassin et al., 1986; Trovero et al., 1991).

Since electrolytic lesions of the VTA spare the cortical NE innervation, whereas 6-OHDA VTA lesions destroy ascending NE fibers that pass near the DA cells, the destruction of the cortical NE innervation could be responsible for this lack of denervation supersensitivity of cortical D1 receptors. A good correlation was found between the extent of damage to NE fibers and the reduction of cortical D1 receptor supersensitivity as compared to that occurring in electrolytically lesioned rats (Tassin et al., 1982a,b).

This permissive role of ascending NE fibers on the appearance of denervation supersensitivity of cortical D1 receptors was also supported by experiments conducted on rats with simultaneous bilateral electrolytic lesions of the VTA and bilateral 6-OHDA lesions of the dorsal NE bundle made near the P.C.S. (Tassin et al., 1986). No significant change in DA-sensitive adenylate cyclase activity was found in the prefrontal cortex of rats with either type of lesion. Lesions of NE neurons alone were also without effect on enzyme activity.

Behavioural consequences of NE/DA interactions

Bilateral electrolytic lesions of the rat VTA induce a condition known as the VTA syndrome, which includes deficits such as locomotor hyperactivity and the disappearance of spontaneous alternation (Le Moal et al., 1976). Correlation studies have indicated that the amplitude of locomotor hyperactivity is proportional to the extent of destruction of the DA fibers innervating the prefrontal cortex and, more interestingly, to the development of a D1 receptor supersensitivity in this area (Tassin et al., 1978c, 1982a). Since destruction of the ascending NE pathways down-regulates cortical D1 receptor denervation supersensitivity induced by the electrolytic lesion of the VTA, we decided to also investigate whether the locomotor hyperactivity induced by this electrolytic lesion could be affected by 6-OHDA-induced lesions of the cortical NE innervation.

In experiments conducted in collaboration with Drs Taghzouti and Simon, animals were either lesioned bilaterally via: 1) electrocoagulations of the VTA; 2) bilateral injections of 6-OHDA into the P.C.S. laterally (site of ascending NE fibers); or 3) simultaneous application of both types of lesions. These experiments revealed that the NE neurons play a permissive role in the expression of the behavioural deficits induced by the electrocoagulation of the VTA. Indeed, while animals with only VTA lesions exhibited a 56% increase in nocturnal locomotor activity and a reduction in spontaneous alternation behaviour (59% instead of 76% for the lesioned and sham-operated animals, respectively), no significant changes in locomotor activity or in spontaneous alternations were seen in rats with simultaneous bilateral electrolytic lesion of the VTA and bilateral 6-OHDA lesions of the dorsal NE bundle when compared to sham-operated rats (Taghzouti et al., 1988). Chemical lesions of the NE fibers alone had no effect either on the locomotor activity or on spontaneous alternation.

Therefore, the simultaneous destruction of NE neurons and of VTA DA neurons markedly reduced the deficits observed in rats with electrolytic VTA lesions. This functional recovery indicates that deficits induced by a given lesion can be abolished by another type of lesion and provides new insights into the antagonistic properties of ascending NE and DA neurons. It is proposed that a functional hierarchy exists between these systems, since no significant modification of locomotor activity or spontaneous alternation was observed in rats with NE lesions alone. These results may explain why many deficits characteristic of the VTA syndrome are more pronounced in rats with electrolytic lesions than in animals with 6-OHDA lesions (Le Moal et al., 1976; Galey et al., 1977; Tassin et al., 1978c).

Since the modulation of D1 receptors by ascending NE neurons seems to occur in the prefrontal cortex, this structure could be an important site of interaction between NE fibers and the target cells of the mesocortical DA neurons involved in the expression of these behavioural deficits. One cannot exclude that this type of interaction could also occur in other structures innervated by VTA DA cells, such as the septum, nucleus accumbens or amygdala.

Opposing effects of NE and DA ascending systems projecting into the prefrontal cortex have also been demonstrated with respect to the anxiolytic-like effects produced by lesions of the prefrontal DA system (Ravard et al., 1990). Injections of 6-OHDA into the prefrontal cortex in the presence of desmethylimipramine, a procedure which destroys DA innervation without affecting the NE nerve terminals, significantly increased punished drinking in saline-pretreated animals. These effects were no longer observed when both prefrontocortical NE and DA fibers were destroyed by a 6-OHDA injection in the absence of desmethylimipramine.

Modulation by alpha-1 adrenergic receptors of the cortical DA -D1- transmission

Locomotor activity may be used, under certain conditions, as an index of cortical D1 transmission

It is generally agreed that the locomotor activating effects induced by the injection of amphetamine into the nucleus accumbens are related to an increased release of DA in this structure (Kelly et al., 1975). As previously mentioned, correlational studies obtained on animals rendered hyperactive by an electrolytic lesion of the VTA indicated that the cortical DA innervation may have an inhibitory role on locomotor activity. However, this locomotor hyperactivity could not be obtained in animals depleted only of cortical DA, suggesting that a modification of DA transmission in the nucleus accumbens is necessary. To test this hypothesis, rats were implanted with chronic bilateral cannulae in both the prefrontal cortex and nucleus accumbens. As expected, injection of amphetamine (1.5 µg/0.5 µl/side) into the nucleus accumbens increased locomotor activity by an average of 240% compared to control saline injections, whereas injection of amphetamine (2.5 μg/0.5 μl/side) into the prefrontal cortex did not alter locomotor activity levels. However, when amphetamine was injected into both sites, amphetamine injected into the prefrontal cortex caused a 42% reduction in the amount of hyperactivity produced when compared to injections into the nucleus accumbens alone (Vézina et al., 1991).

Additional experiments have shown that this cortical effect is mediated via D1 receptors. The injection of SCH 23390 (0.25 μg/0.5 μl/side), a D1 antagonist, into the prefrontal cortex causes a 110% increase in the locomotor activity induced by injection of amphetamine into the nucleus accumbens, whereas the injection of sulpiride (1 μg/0.5 μl/side), a D2 antagonist, is without significant effect. Both of these antagonists, at the above doses, blocked the locomotor activating effects of amphetamine when co-injected with the drug into the nucleus accumbens (Vézina et al., 1991).

These results suggest that the inhibitory role of prefrontocortical DA innervation on locomotor behaviour is mediated by D1 receptors. Modifications in locomotor activity may therefore reflect changes in D1 cortical transmission if, and only if, DA transmission in the nucleus accumbens has first been either inhibited (electrolytic lesion of the VTA) or facilitated (local amphetamine injection).

Blockade of alpha-1 adrenergic receptors reverses locomotor hyperactivity induced by electrolytic lesion of the VTA

Since the destruction of ascending NE neurons was able to compensate for the behavioural deficits induced by the electrolytic lesion of the VTA, we decided to see whether the pharmacological blockade of either the alpha or β NE receptors could reverse these deficits. This was carried out in behavioural experiments in which animals, rendered hyperactive by electrolytic lesions of the VTA, were treated with low doses (0.5 mg/kg, i.p.) of prazosin (alpha-1 antagonist), yohimbine (1.6 mg/kg, i.p., alpha-2 antagonist) or propranolol (5 mg/kg, i.p., β antagonist), one hour before measuring their level of nocturnal locomotor activity. These doses were chosen since they were the highest doses which had no effect on the activity of sham-operated animals tested on the night following the injection. Prazosin completely abolished the locomotor hyperactivity seen in lesioned animals for periods lasting up to 24 hours following the injection, whereas no significant effects on locomotor activity could be observed when animals were tested in the presence of either yohimbine or propranolol. The effect of prazosin seemed to be specific for a particular subtype of alpha 1-adrenergic receptor since the same experiment performed with WB 4101, another alpha-1 adrenergic antagonist which binds to alpha-1A and alpha-1B adrenergic receptors (Morrow and Creese, 1986), did not modify the locomotor hyperactivity of lesioned animals. Indeed, autoradiographic experiments using tritiated prazosin and WB 4101 as ligands have confirmed the presence of two different patterns in the cortical binding sites of these ligands (Trovero et al., 1992b).

Cortical alpha 1-adrenergic receptors sensitive to prazosin modulate prefrontocortical D1 transmission

As previously outlined, the effects of NE denervation on the behavioural deficits induced by electrolytic lesions of the VTA could involve any of the forebrain structures innervated by DA fibers. This is also valid for the results obtained with intraperitoneally injected prazosin. Whether these results were dependent on actions in the prefrontal cortex was therefore tested in animals injected simultaneously in the prefrontal cortex and the nucleus accumbens (see above). The injection of prazosin (0.6 nmole/ 0.5 µl/side) into the prefrontal cortex did not modify the locomotor activity induced by injecting amphetamine (1.5 µg/0.5 µl/side) bilaterally into the nucleus accumbens when compared to saline-treated rats. However, a similar injection of prazosin into the prefrontal cortex completely reversed the locomotor hyperactivity of animals simultaneously injected with amphetamine into the nucleus accumbens (Blanc et al., in preparation), suggesting a facilitatory role of prazosin on cortical D1 transmission.

Involvement of alpha-1 adrenergic receptors in the development of an EEDQ-induced hypersensitivity of DA-sensitive adenylate cyclase (D1 receptors) in the prefrontal cortex

N-ethoxy-2-ethoxy-1, 2 dihydroquinoleine (EEDQ) is an irreversible pharmacologic antagonist at a variety of neurotransmitter receptors including dopaminergic (D1, D2), serotonergic (5-HT2) and noradrenergic (alpha-1, alpha-2) receptors (Meller et al., 1985), and peripheral administration of EEDQ results in a dose-dependent reduction in the density of these receptors. EEDQ treatment has been employed to elucidate the stoichiometric relationship of the coupling of neurotransmitter receptors with their second messenger systems. For example, in the striatum, a large (40%) proportion of D1 binding sites could be inactivated by EEDQ treatment without any significant reduction in the maximal activity of DA-stimulated adenylate cyclase (Hess et al., 1987), indicating the presence of a population of "spare" D1 receptors not coupled to transducing subunits.

Since EEDQ can be used as a tool to elucidate the mechanisms of D1 receptor/effector coupling, we have studied the consequences of an in vivo EEDQ treatment on D1 site density and DA-sensitive adenylate cyclase activity in the prefrontal cortex. Surprisingly, we found that a small reduction in cortical D1 receptor density, induced by a low dose of EEDQ (0.8 mg/kg, i.p.) and visualized by autoradiography was accompanied four

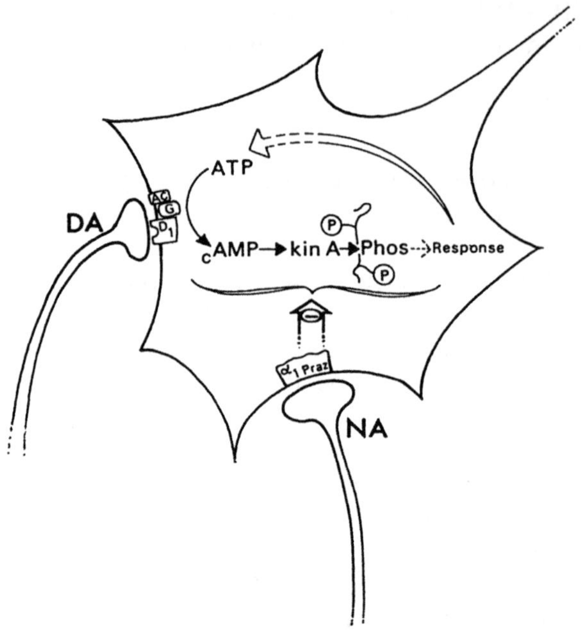

Fig. 1. Illustration of the role of alpha-1 adrenergic receptors on cortical D1 transmission. In the rat prefrontal cortex, the destruction of NE ascending fibers or the blockade by prazosin of a possibly specific alpha-1 adrenergic receptor compensate for behavioural deficits induced by electrolytic lesions of the VTA-DA cell bodies. Moreover, a partial irreversible inactivation of D1 receptors by EEDQ induces a compensatory hyperactivity of cortical DA-sensitive adenylate cyclase, which is blocked by pretreatment of the animals by prazosin

hours later, by an increase in cortical DA-sensitive adenylate cyclase activity. These paradoxical effects were not found in the striatum of the same EEDQ-treated animals where, although there was also a small reduction in striatal D1 receptor density, DA-sensitive adenylate cyclase activity was unchanged.

Our previous data indicate that modulation of cortical D1 receptor density is at least partly controlled by NE neurons via stimulation of alpha-1 adrenergic receptors sensitive to prazosin. The autoradiographic analysis of cortical slices of EEDQ-treated animals revealed that rats with a hypersensitivity of cortical D1 receptors were those with less than 35% of their cortical alpha-1 adrenergic receptors inactivated. Rats were therefore injected with prazosin (1.5 mg/kg, i.p.) 30 minutes before and 2 hours after (1.5 mg/kg, i.p.) the administration of the same low doses of EEDQ (0.8 mg/kg, i.p.), in order to block alpha-1 adrenergic receptors. This treatment did not modify the decrease in 3H SCH 23390 binding to cortical D1 receptors but blocked the increased activity of the cortical DA-sensitive adenylate cyclase (Trovero et al., 1992b).

Figure 1 illustrates our findings on the interactions between alpha-1 adrenergic receptors and cortical D1 transmission. Although there is no direct evidence that cortical alpha-1 adrenergic and D1 receptors are located on the same cells, this possibility cannot be excluded. The electrophysiological studies of Bunney and Aghajanian (1976) would favor this hypothesis. The presence of 3H-prazosin and 3H-SCH 23390 binding sites respectively in layers III and V/VI of the prefrontal cortex indicates that, if these receptors are on the same cells, they must be located on different regions (e.g., soma, dendrite, axon) of the target neurons.

Unlike the results of Magistretti and Schorderet (1985), who have shown a potentiation by alpha-1 adrenergic receptors of cyclic AMP accumulation elicited by vasoactive intestinal polypeptide in the mouse cerebral cortex, our data indicate that the message induced by the stimulation of cortical D1 receptors is inhibited by alpha-1 adrenergic receptors. Indeed, it can be assumed that the hypersensitivity of DA-sensitive adenylate cyclase induced by EEDQ is the result of a positive feed-back aimed at restoring cortical D1 transmission. The absence of EEDQ-induced hypersensitivity of DA-sensitive adenylate cyclase following the blockade of the alpha-1 adrenergic receptors by prazosin suggests that prazosin facilitates cortical D1 transmission.

Three main conclusions can be drawn from our data:

First, the collateral sprouting of cortical DA nerve terminals, induced by the disappearance of NE axons, indicates that DA fibers are in physical competition with NE axons in structures where both types of innervation co-exist.

Secondly, NE and DA neurons can modulate DA and NE receptor sensitivities, respectively. Although these hetero-regulations have been demonstrated only for D1-DA and β adrenergic receptors, they may, however, also occur at other NE or DA receptor subtypes.

Thirdly, the regulation of cortical DA (D1) receptors by NE neurons may explain why different behavioural deficits induced by cortical DA denervation (e.g., locomotor hyperactivity, disappearance of spontaneous alternation and diminished reactions to anxiogenic situations), are counterbalanced by superimposed NE lesions. Indeed, experiments have indicated that NE neurons, via the stimulation by NE of a specific type of prazosin-sensitive alpha-1 adrenergic receptors, may act on cells bearing DA receptors and modulate cortical D1 transmission.

Our data provide evidence that NE and DA neurons innervating the prefrontal cortex have antagonistic properties on the behavioural outputs they modulate. A functional hierarchy between these two systems appears to exist, since no behavioural modification is observed when only the NE neurons are lesioned. It should be mentioned, however, that Arnsten and Goldman-Rakic (1985) have presented data suggesting that NE and DA transmissions act in synergy. Indeed, these authors have shown that cognitive deficits due to aging and possibly related to a loss of cortical DA can be compensated for by treating animals with clonidine, an alpha-2 adrenergic receptor agonist. These two conclusions could be reconciled if it is assumed that, under their experimental conditions, the stimulation of presynaptic alpha-2 adrenergic receptors by clonidine has decreased NE release. In any case, the putative roles of NE and DA neurons in the pathogenesis of various mental diseases compel us to take into consideration interactions between neuromodulatory systems.

Possible implications of neuromodulators in schizophrenia

Similarities and differences among the neuromodulatory neuronal systems

The concept that specific neurons modulate information processing rather than convey sensory or motor signals seems to be well established (see, for example, Servan-Schreiber et al., 1990). In any representation of the primary pathways responsible for the processing of sensory stimuli or motor outputs (Nauta and Feirtag, 1979), it is notable that these pathways probably include neurons releasing GABA and excitatory amino-acids, and possibly acetylcholine or neuropeptides, whereas NE, DA or 5-HT neurons do not appear to be involved (Fig. 2).

The absence of this latter group of neurons in the main communication pathways of the central nervous system is both logical and surprising. It is logical if we consider that the entire group of monoaminergic cells represents less than 1% of the total brain neurons, and surprising, since therapeutic improvements obtained with most psychotropic drugs, such as neuroleptics and antidepressants, appear to act via modification of monoaminergic neurotransmission.

In fact, monoaminergic neurons have an anatomical distribution that is clearly different from that of other neurons which connect brain structures (Fig. 3). Their cell bodies are gathered in the mesencephalon (ventral tegmental area, substantia nigra, locus coeruleus and anterior raphe nuclei) and send divergent axons toward different forebrain areas. These axons are not myelinated and, therefore, have relatively slow conduction velocities ($<0.8\,m/s$).

The anatomical divergence of these neurons suggests that their networks are superimposed on those of neurons conveying sensory or motor signals. Monoaminergic neurons would favour the functional role of one or another cerebral structure, depending on the behavioural and external situations the animal is exposed to. They would create a functional hierarchy between structures and allow the performance of different tasks within the same basic neuronal networks.

Although monoaminergic neurons exhibit similar features, they possess clear anatomo-functional differences:

First, on anatomical grounds, even if DA neurons expand their territories of innervation in the course of phylogeny, particularly in premotor and motor cortices (Berger et al., 1986; Lewis et al., 1987), they are less widely distributed than are NE and 5-HT neurons. Moreover, the specific destruction of either of these two latter neuronal systems in the rat induces

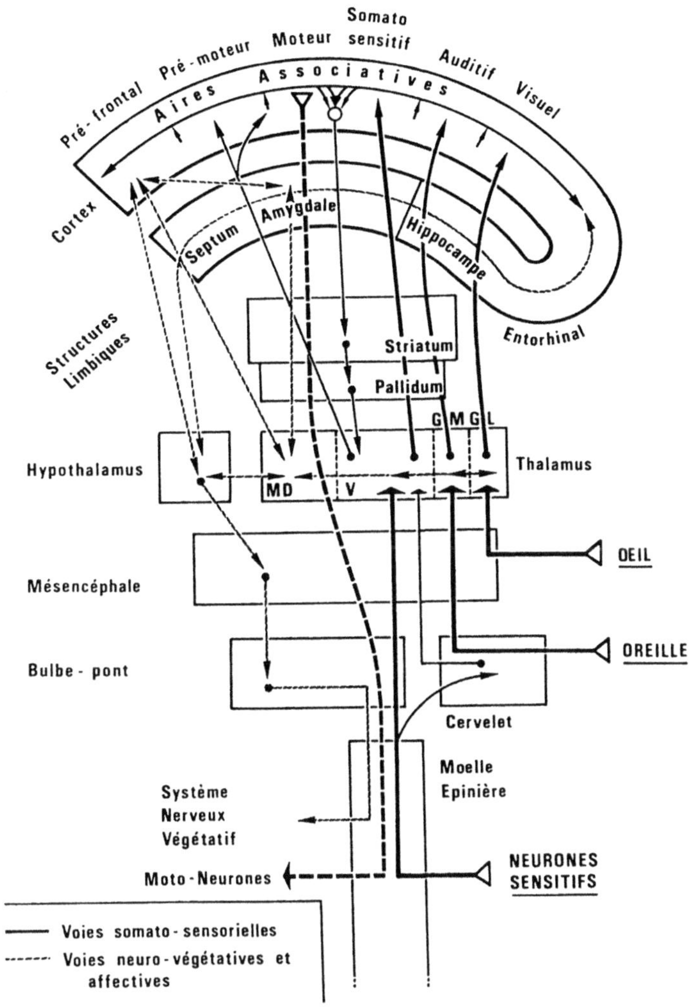

Fig. 2. Main communication pathways in the central nervous system. Each sensory message (auditory, visual or tactile) is first processed by a specific thalamus nucleus [medial geniculate (*GM*), lateral geniculate (*GL*), or ventral (*V*)]. Following their reception in the cortical primary sensory areas, sensory modalities are associated up to the ultimate associative area, the prefrontal cortex. The prefontal cortex is connected with most of the limbic structures, the hypothalamus, the medio-dorsal nucleus of the thalamus (*MD*) and to subcortical structures via the cortico-striato-pallido-thalamo-cortical loop. NE, DA and 5-HT neurons are not part of these main communication pathways

behavioural deficits which are far less severe than those induced by denervation of ascending DA neurons. Apart from the studies by Mason (1980), there is no indication of behavioural deficits induced by depletion of NE ascending bundles. Destruction of 5-HT neurons makes animals impulsive and aggressive (Soubrié, 1986), while lesion of DA ascending neurons disturbs most of the brain integrative functions not directly linked to the

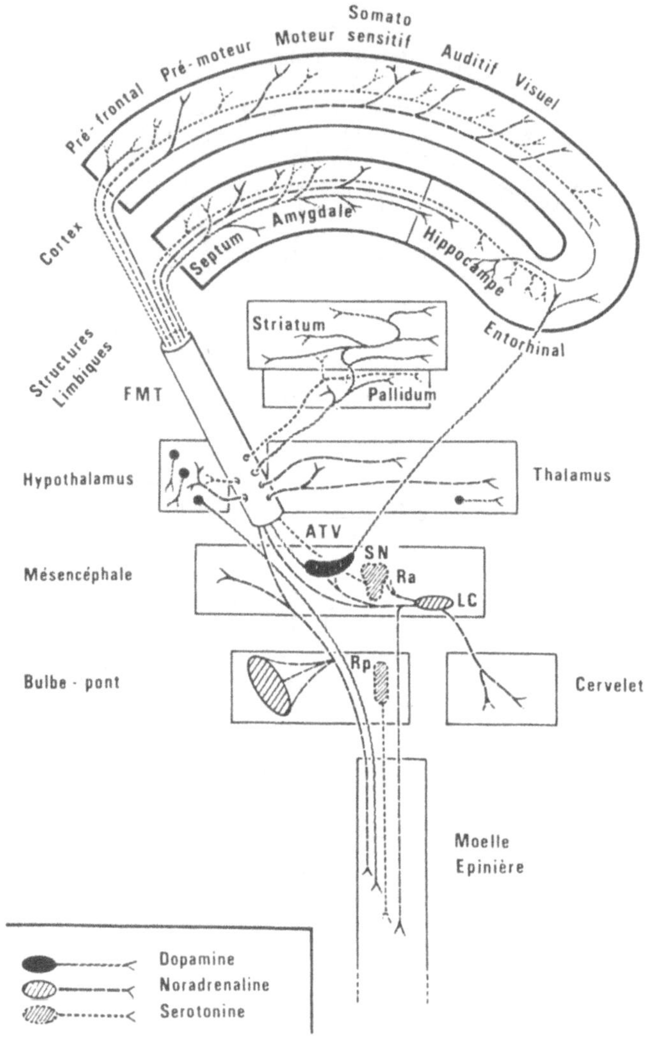

Fig. 3. Neuromodulators circuitry. Cell bodies of NE, DA and 5-HT neurons, located in the mesencephalon [locus coeruleus (*LC*), ventral tegmental area (*ATV*) and substantia nigra (*SN*), and anterior raphe (*Ra*)] send divergent axons towards forebrain structures. This anatomical characteristic is part of the evidence suggesting that these neurons play a neuromodulatory role in the central nervous system. *Rp* is the abbreviation for posterior raphe nuclei. *FMT* medical forebrain bundle

sensory and motor related processes, such as learning and memory, cognitive functions and reinforcement processes (Simon and Le Moal, 1991). In the absence of DA neurons animals have a total behavioural disorganization and an inability to initiate and adapt.

Reactivities to sensory stimuli are also different:

Non-specific visual or auditory stimuli lead to an excitation, followed by an inhibition of the unit activity of 5-HT cells of the anterior — dorsal and centralis superior — raphe nuclei (Heym et al., 1982; Rasmussen et al., 1984). There is no habituation, since the repetition of identical stimulations will still trigger the same neuronal responses.

In freely moving animals, NE cells of the locus coeruleus also repond to a variety of stimuli, including auditory, visual and tactile, with an excitation-inhibition pattern (Aston-Jones and Bloom, 1981a). However, unlike 5-HT neurons, the response rapidly habituates with repetitive stimuli, with the desensitization developing in parallel with the animal's disinterest in the stimulus.

Repetitive presentation of auditory clicks or flashes elicits a similar pattern of excitation followed by inhibition in most DA neurons, with no evidence of habituation (Steinfels et al., 1983). Surprisingly, although the nigro-striatal DA pathway is considered to play a role in the control of motor function, the activity of nigral DA neurons is not changed by movement per se. Modifications of the activity of these neurons seem to be related to motivational arousal elicited by stimuli with appetitive properties (Romo and Schultz, 1990). For example, opening the door of the room in which the cat's chamber is housed will trigger an orienting response, associated with a short latency excitation followed by a long-lasting depression of DA unit activity. The depression corresponds to the period of fixation and focussed attention (Steinfels et al., 1983; Jacobs, 1985). There is some evidence that this period of focussed attention is related to the activation of the DA neurons innervating the prefrontal cortex and may reflect the existence of interactions between ascending mesocortical and mesosubcortical DA neurons (Lavielle et al., 1979; Carter and Pycock, 1980; Tassin et al., 1982; Simon and Le Moal, 1988). Interestingly, when stimuli having behavioral significance for the animal are presented repetitively, a habituation appears for the DA neurons in a manner similar to that observed for NE neurons following repeated non specific stimulations.

Post-synaptic events due to release of these neuromodulators are also different:

It is well known that NE decreases the spontaneous activity of target cells without changing the amplitude of evoked responses, the final effect being an increase of the signal-to-noise ratio (Aston-Jones and Bloom, 1981a). Some experiments have also indicated that NE could potentiate responses to both excitatory and inhibitory inputs with only minimal change in background firing rate (Woodward et al., 1979). However, 5-HT, at least in the primary cortical somatosensory areas, decreases the signal-to-noise ratio by inhibiting the spontaneous cell activity to a smaller extent than the evoked response (Waterhouse et al., 1986). The post-synaptic effects of DA appear to be more complex: in most cases DA appears to inhibit spontaneous activity as well as the evoked responses of target cells, at least as shown by extracellular recordings (Bunney, 1979). In fact, DA probably modifies the sensitivity of post-synaptic cells to other neurotransmitters through changes in endogenous cyclic AMP levels. For example, in monkeys with Parkinsonism induced by 1-methyl- 4-phenyl-1,2,3,6-tetrahydropyridine

(MPTP), pallidal neurons responded without selectivity; i.e. there was an increase in both the number of responding neurons and the magnitude of the responses following electrical stimulation of the striatum (Tremblay and Filion, 1989; Filion et al., 1989). Some experiments, performed on the retina or the substantia nigra have shown that DA may also act by decreasing electrotonic coupling between target cells (Piccolino et al., 1984; Grace and Bunney, 1983). Finally, a synergism between striatal D1 and D2 receptors was recently shown to induce an inhibition of (Na+ + K+) ATPase (Bertorello et al., 1990).

It is clear that each of these responses are those most commonly induced by monoaminergic neurons and depend upon the subtype of receptor stimulated. Nevertheless, these examples illustrate the main characteristic of these neurons; i.e. the capacity to modulate messages by changing the sensitivity of target cells to other neurotransmitters. Analysis of their functional differences may, however, give some indications about the consequences of their interactions.

Last of the differences that we would like to underline is the modification of the activity of monoaminergic neurons across the sleep-wake cycle. These have been extensively reviewed by Jacobs (1985): 5-HT and NE neuronal activity manifest a high degree of state-dependency, while the activity of DA neurons is virtually state-independent. The activity of 5-HT neurons of the dorsal raphe nucleus as well as that of NE neurons of the locus coeruleus shows a gradual decline as the animal becomes drowsy and enters slow wave sleep. These neurons become virtually silent in REM sleep. NE cells of the locus coeruleus have also been shown to anticipate fluctuations in the sleep-waking cycle (Aston-Jones and Bloom, 1981b).

Surprisingly, DA neurons display a remarkable degree of stability across quiet waking, slow wave sleep and REM sleep (Steinfels et al., 1983). It should be recalled, however, that although these recordings have been performed in the ventral tegmental area and the substantia nigra, respectively the A10 and A9 DA cell groups, there is no clear data on the activities of mesocortical DA neurons across the sleep-wake cycle. DA cell bodies innervating the prefrontal cortex represent only 3% of the total population of A10 DA cells and were probably not recorded in these experiments. We cannot therefore exclude the possibility that mesocortical DA neurons are affected during sleep, since some data have shown that these cells decrease their activities following diminished environmental stimuli such as long-term isolation (Blanc et al., 1980).

Before trying to determine whether these characteristics of monoaminergic neurons may be related to schizophrenia, it is important to note that, although all these data have been obtained on animals, the level of reproducibility from one species to another provides some grounds for assuming that they may be valid for man.

How to approach schizophrenia in basic neurosciences?

Among the different criteria proposed by the DSM III R for the diagnosis of schizophrenia, the presence of prominent hallucinations and of marked loosening of associations seem particularly appropriate for a neurobiological approach. One possible way to understand how hallucinations appear is to find out whether some common dysfunctions can be induced by hallucinogens. Obviously, hallucinogens are a heterogeneous family of drugs, including those, like LSD, that produce essentially visual hallucinations and those like phencyclidine (PCP) and the methoxyamphetamines which induce mainly auditory hallucinations. However, it seems that some common neurochemical correlates can be identified from a rigorous analysis of their actions (Hamon, 1984):

First, after in vivo injections of 3H-LSD in the squirrel monkey (Snyder and Reivich, 1966) and 3H-2,5-dimethoxy-4-methylamphetamine (3H-DOM) in the cat (Idapaan-Heikkila et al., 1969), striking similarities were found in the structures exhibiting the densest labeling. These were the visual and auditory areas (lateral geniculate, superior colliculi, medial geniculate, inferior colliculi) and limbic areas such as the amygdala and the hippocampus, regions involved in the transfer (geniculate bodies) and processing (limbic areas) of sensory messages.

Second, different authors (Meibach et al., 1979; Crosby et al., 1982; Hammer et al., 1982) have used the 3H-2-deoxyglucose technique, developed by Sokoloff et al. (1977), to note both increases and decreases in glucose utilization in various brain areas following the administration of PCP or ketamine to rats. A general *increase* in metabolism was observed in corticolimbic areas, such as part of the hippocampus, the frontal cortex and the cingulate gyrus. In contrast, a *decrease* in metabolism was detected in inferior colliculi and auditory and sensorimotor cortex.

These neurochemical data indicate that hallucinations evoked by hallucinogens could be due to the induction by these drugs of a functional imbalance between limbic and sensory areas. The association between the decreased metabolic activity of sensory structures, which would diminish the perception of external information, and the increased activity of limbic areas, facilitating the emergence of endogenous messages, would be responsible for hallucinations. A change in this sensory/limbic balance may, in fact, be partly explained by the action of hallucinogens on neuromodulator transmission. We would like to keep the putative existence of this functional balance between limbic and sensory areas as a guide line in the following discussion.

There is a considerable amount of biochemical and electrophysiological work which has shown that hallucinogens decrease the activity of anterior raphe nuclei 5-HT cells and 5-HT utilization in forebrain structures. This has been shown for LSD (Freedman, 1961; Rosecrans et al., 1967; Haigler and Aghajanian, 1977), DMT and psilocybin (Anden et al., 1971), p-methoxyamphetamine and 2,5-dimethoxy-4-methylamphetamine (DOM)

(Anden et al., 1974; Leonard, 1973), PCP or ketamine (Johnson, 1982; Martin and Smith, 1982) and delta 9 tetrahydrocannabinol (Sofia et al., 1971). Even if this reduced turnover of 5-HT seems critical in the action of hallucinogens, it is obvious, however, that all drugs which reduce 5-HT turnover in brain are not hallucinogenic. Other correlates indicate that most hallucinogens share the common property of activating catecholamine turn over. LSD increases DA turnover in striatal, limbic and prefrontocortical areas (Persson, 1978; Bowers and Salomonsson, 1982). This drug has, moreover, been described as a partial agonist of DA-D1 receptors (Von Hungen et al., 1975; Bockaert et al., 1976) and Freedman (1963) has shown that an acute injection of LSD decreased cerebral NE levels. This was later related to an increased NE turnover (Peters and Tang, 1977). Other hallucinogens accelerate catecholamine turnover. Psylocibin and DMT (Anden et al., 1971) as well as DOM (Leonard, 1973) and tetrahydrocannabinol (Bloom, 1982) increase NE turnover. PCP (Vickroy and Johnson, 1982; Howard-Butcher and Levine, 1982) and tetrahydrocannabinol (Bloom, 1982) induce clear activations of DA transmission or turnover.

Data from the literature suggest that hallucinogens may evoke hallucinations because they both modify a hypothetic sensory/limbic balance in the brain and change the equilibrium between the three main neuromodulatory transmissions; they induce a reduction in 5-HT turnover associated with an increase in catecholamine utilization.

It may be interesting to compare the effects produced by hallucinogens on the brain with those occurring during sleep and more precisely during REM. As previously mentioned, this particular state of brain functioning is associated with an almost total disappearance of NE and 5-HT cell activities, whereas at least the population of DA neurons projecting to subcortical structures is unaffected. Moreover, since the processing of external stimuli is inhibited during REM, it is likely that, if we refer to our proposition of sensory/limbic balance, this equilibrium is displaced towards the limbic functioning as it is following hallucinogen ingestion. The main difference between the effects of hallucinogens and REM on neuromodulatory neurons would be that NE neurons, which are silent in REM, are still active and even activated by hallucinogens.

If we now try to analyze the characteristics of brain function during REM, it is clear that one of its most striking properties during this period is the capacity to construct images. They are structured in space in such a precise way that the dreamer may, for example, feel that he is walking under a lane of trees whose leaves are agitated by the wind. The second interesting point is that the images rarely stay immobile and are quickly associated one after the other, the dreamer being driven in an uncontrolled adventure. Besides these capacities, which would be exceptional in an awake state, some functions seem impossible to fulfil during a dream. For example, even if we can feel that we are reading a book by simulating printed pages, attempts to read these pages are in vain.

These observations suggest that the brain is able to operate in another mode to that is selected to perform cognitive tasks. In this mode, neuronal networks would communicate following an analogical process where activated neurons could switch from one network to another (Hooper and Moulins, 1989) and induce the activation of a neighboring network. Such communications between networks, which are probably much quicker than those necessary for the performance of cognitive tasks, would not be at random. However, they would not produce associations of the same type as those obtained when neuromodulators are normally activated by external stimuli. It is possible that the "marked loosening of associations", depicted in the DSM III R, is related to this mode of brain functioning. Similarly, the psychoanalytical subconscious described by Freud (Freud, 1915; Laplanche and Pontalis, 1968) could be supported, at least partly, by the existence of such associations (Tassin, 1989). In any case, the spectacular achievements performed by some psychotic patients such as the twins described by Sachs (1988), may be due to their pathological possibility to make the brain work continuously on this mode in an awake state: these twins were able to evaluate precisely, in few seconds, the presence of 111 matches on the floor without counting them.

Respective roles of neuromodulators in these two modes of brain functioning

If a person, while asleep, detects an external stimulus sufficiently intense or significant to prevent him from integrating it into a dream, NE and 5-HT neurons are activated and awakening occurs, inducing a switch in the mode of brain functioning. As previously proposed, it is also possible that mesocortical DA neurons are also stimulated. This arousal corresponds to a modification of the sensory/limbic brain balance and allows cognitive tasks to be performed (Fig. 4).

Up to now, most of the neurochemical data concerning the cognitive processes have been focussed on the role of mesocortical DA neurons, cither in the monkey (Brozowski et al., 1979) or in the rat (Simon et al., 1980). More recently, Sawagachi and Goldman-Rakic (1991) have emphasized the role of cortical D1 transmission in working memory and these findings are in good agreement with our own studies (Vézina et al., 1991; this chapter, see above). Moreover, stimulation of a specific type of cortical alpha 1-adrenergic receptor seems to inhibit cortical D1 transmission, suggesting that an activation of NE neurons will temporarily block a cognitive treatment of external stimuli to allow the brain to process information in its analogical mode of functioning, i.e. when the sensory/limbic balance favours the metabolic activities of limbic structures. Rapid habituation of the locus coeruleus NE neurons to the activation induced by repeated external stimuli, suggests that this modification of brain processing is essentially linked to the treatments of novel stimuli. This proposition implies

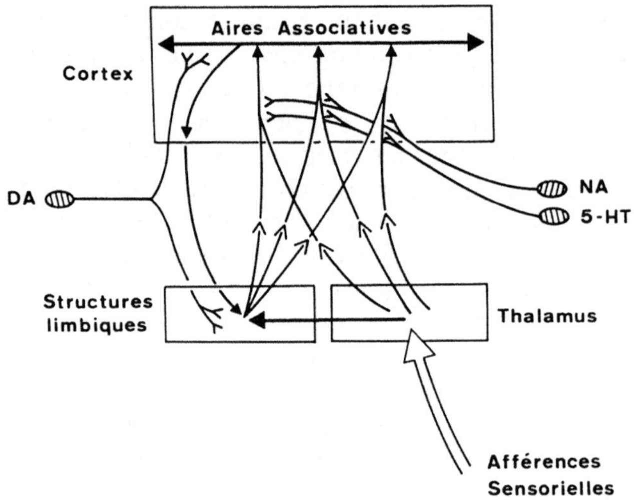

Fig. 4. Schematic representation of the two possible modes of brain function. Sensory afferents reaching the thalamus may be processed either in an analogical or a cognitive mode depending on the respective activities of NE, 5-HT and cortical or subcortical DA neurons. These activities are under the control of external non-specific (NE and 5-HT neurons) or signifying (DA neurons) stimuli. Cognitive mode is characterized by regular activities of anterior raphe nuclei 5-HT and mesocortical DA neurons. An activation of NE neurons, which inhibit cortical D1 transmission, will switch the brain processing towards the analogical mode. DA neurons control the functional hierarchy between cortical and subcortical structures

that, while awake, both modes of functioning may occur sequentially. The impossibility of reading sentences during REM may be related to the incapacity of the brain to switch from one mode to the other in that period.

The presence of a putative bimodal processing of information has already been suggested by observations showing that the visual capacity of neurological patients can be dissociated between the perceptual report of a visual stimulus and the ability to direct spatially accurate movements towards that stimulus (see references in Goodale et al., 1991). Similarly, Freudian slips or failed acts can be considered as expressions of the interference of both brain modes in an awake state.

These two modes of brain functioning could be described as analogical and cognitive.

Results obtained with hallucinogens suggest that activation of anterior raphe nuclei 5-HT neurons is necessary to avoid hallucinations and to allow the brain to process correctly external perceptions. Moreover, the resistance of these neurons to habituation suggests that their role is particularly critical in a "routine" arousal. The action of 5-HT neurons on D1 cortical transmission would be, in some way, opposite to that of NE neurons. In line with this hypothesis, we have recently found that rats depleted in forebrain 5-HT by 5,7-DHT injections in the anterior raphe nuclei have a diminished activity of cortical DA-sensitive adenylate cyclase (Trovero et al., unpublished observations).

Schizophrenia and neuromodulators

The evidence that we have presented suggests that hallucinations and marked loosening of associations, two characteristic symptoms of schizophrenia, may be explained if we assume that, in this mental disease, the sequential oscillations of the proposed sensory/limbic brain balance are not correctly linked together during wakefulness. Our proposition would be that schizophrenic patients suffer from an abnormal preponderance of the analogical mode of brain processing.

With regard to neuromodulators, the schizophrenic state could be characterized by one or more of these dysregulations:

1. An increased sensitivity of NE neurons to environmental stimuli, possibly associated with a disappearance of the habituation of these neurons to repeated stimulations.
2. A decreased sensitivity of anterior raphe nuclei 5-HT neurons to environmental stimuli. It cannot be excluded that 5-HT neurons, in this pathological situation, become desensitized by repeated stimulations.
3. Mesencephalic DA neurons, whose role seems to create a functional hierarchy between cortical and subcortical structures, may be responsible both for a DA overactivation of subcortical structures and for a decreased DA stimulation in the prefrontal cortex. This latter possibility, which is a correlate of previously proposed interactions between cortical and subcortical DA systems (Carter and Pycock, 1980; Tassin et al., 1982b, 1991; Glowinski et al., 1984; Weinberger, 1986; Simon and Le Moal, 1988; Hervé et al., 1989), may explain the presence of a limited prefrontal physiological reserve in some schizophrenic patients (Berman et al., 1986).

Because of interactions, including those due to receptor heteroregulation, between the three neuromodulatory systems, it is likely that the presence of conditions 1 or 2 will, with time, induce 3.

If our hypothesis is correct, it might give some indications about the origins of the illness. At birth, sensory receptors are mature; the auditory system is even functional before birth (Lecanuet et al., 1987). Since cortical myelinized associative pathways and neuromodulatory systems are not yet mature, the preponderant mode of brain functioning would be the analogical one, i.e. a mode which is incompatible with the cognitive process of information. This is probably why adults are not able, at will, to recall events from their first two years of life. These first sensations have however been stored and, most importantly, they are likely to play a critical role in the early development of the reactivity of mesencephalic DA neurons to "signifying" — although not processed in the cognitive mode — stimuli. If we compare it with the cognitive mode which is, by nature, permanently confronted to reality, one of the characteristics of the analogical mode is its sensitivity to errors. Such putative "errors" in the processing of signifying environmental stimuli may then become particularly handicapping at adolescence, a period where neuromodulatory systems increase their activities and schizophrenia usually occurs.

Dysregulations of NE and 5-HT neurons may also be determinant in this mental disease. However, the mechanisms controlling the activities of these neuromodulatory neurons, especially those explaining why NE cells habituate and 5-HT cells do not, are still poorly understood. The presence of autoreceptors on both types of cells, the alpha-2 adrenergic and 5-HT1A types for NE and 5-HT cells respectively, does not explain the differences observed. Basic biological deficits in the regulation of the activity of either NE or 5-HT cells in schizophrenic patients cannot therefore be excluded.

Concluding remarks

In the first part of this presentation, we have described some data suggesting the presence of interactions between NE and DA neuronal systems in the rat prefrontal cortex. Besides the physical interaction which occurs between NE and DA axons, as shown by the sprouting of DA nerve terminals following the depletion of NE ascending fibers, indications of reciprocal heterogulations between NE and DA receptors were reviewed. Among these, results suggesting an inhibitory role on cortical D1 transmission by NE stimulation of a possibly specific alpha-1 adrenergic receptor, may have clinical implications.

In the second part, we have tried to examine the likely influence of neuromodulatory systems in schizophrenia. The analysis of modifications in brain metabolism induced by hallucinogens has led us to propose the existence of a sensory/limbic balance characterized by increased activities of limbic structures during the course of hallucinations. Neurochemical correlates indicate that hallucinogens induce a decreased activity of anterior raphe nuclei 5-HT neurons whereas catecholaminergic neuronal activities are increased. These data have been compared with those obtained during REM where it is postulated that the sensory/limbic balance is in favor of limbic structures and that the brain works in an analogical mode.

While awake, achievements of cognitive tasks necessitate another mode of information processing. This cognitive mode would be characterized by regular activations of both anterior raphe nuclei 5-HT cells and mesocortical DA neurons. The intermittent activation of locus coeruleus NE cells, which has been shown to inhibit the cortical D1 transmission, would switch the brain from the cognitive to the analogical mode. It is possible that the therapeutic effects of some neuroleptics are partly linked to their blockade of alpha-1 adrenergic receptors. We propose that schizophrenic patients with "positive" symptoms suffer, during wakefulness, from an abnormal preponderance of the analogical mode of brain processing. Conversely, "negative" symptoms may be due to a hyporeactivity of NE neurons.

If these hypotheses are correct, further experiments should demonstrate that mesocortical DA neurons are, unlike DA neurons innervating subcortical structures, affected by sleep-wake cycle fluctuations. Finally, a better knowledge of the biological mechanisms controlling the reactivities of NE and 5-HT neurons, especially the reason why the former habituate and the latter do not, may help to understand the etiopathogeny of schizophrenia. Pure biological deficits can be responsible for dysregulations of monoaminergic activities. However, it cannot be excluded that specific and repeated variations in the early environment of the patients have induced an inability to discriminate between novel and usual stimuli, a critical factor to settle regulations of monoaminergic neurons.

References

Anden NE, Corrodi H, Fuxe K (1971) Hallucinogenic drugs of the indolealkylamine type and central monoamine neurons. J Pharmacol Exp Ther 179: 236–249

Anden NE, Corrodi H, Fuxe K, Meck JL (1974) Hallucinogenic phenylethylamines interactions with serotonin turnover and receptors. Eur J Pharmacol 25: 176–184

Antelman SM, Caggiula AR (1977) Norepinephrine-dopamine-interactions and behavior. Science 195: 646–653

Arnsten AFT, Goldman-Rakic PS (1985) Alpha 2-adrenergic mechanisms in prefrontal cortex associated with cognitive decline in aged nonhuman primates. Science 230: 1273–1276

Aston-Jones G, Bloom FE (1981a) NE-containing locus coeruleus neurons in behaving rats exhibit pronounced responses to non-noxious environmental stimuli. J Neurosci 1: 887–900

Aston-Jones G, Bloom FE (1981b) Activity of norepinephrine containing Locus coeruleus neurons in behaving rats anticipates fluctuations in the sleep-waking cycle. J Neurosci 1: 876–886

Berger B, Tassin JP, Blanc G, Moyne MA, Thierry AM (1974) Histochemical confirmation for dopaminergic innervation of the rat cerebral cortex after destruction of NA ascending pathways. Brain Res 81: 332–337

Berger B, Trottier S, Gaspar P, et al (1986) Major dopamine innervation of the cortical motor areas in the Cynomolgus monkey. A radioautographic study with comparative assessment of serotonergic afferents. Neurosci Lett 72: 121–127

Berman KF, Zec RF, Weinberger DR (1986) Physiological dysfunction of dorsolateral prefrontal cortex in schizophrenia. II. Role of medication, attention, and mental effort. Arch Gen Psychiatry 43: 126–143

Bertorello AM, Hopfield JF, Aperia A, Greengard P (1990) Inhibition by dopamine of (Na+ + K+) ATPase activity in neostriatal neurons through D1 and D2 dopamine receptor synergism. Nature 347: 386–388

Björklund A, Lindvall O (1978) The meso-telencephalic dopamine neuron system: a review of its anatomy. In: Livingstone KE, Hornykiewicz O (eds) Limbic mechanisms. Plenum Press, New York

Blanc G, Hervé D, Simon H, Lisoprawski A, Glowinski J, Tassin JP (1980) Response to stress of meso-cortico-frontal DA neurons in rats after long-term isolation. Nature 284: 265–267

Bloom AS (1982) Effect of delta 9-tetrahydrocannabinol on the synthesis of dopamine and norepinephrine in mouse brain synaptosomes. J Pharmacol Exp Ther 202: 519–531
Bockaërt J, Premont J, Glowinski J, Thierry AM, Tassin JP (1976) Topographical distribution of dopaminergic innervation and of dopaminergic receptors in the rat striatum. II. Distribution and characteristics of dopamine adenylate cyclase. Interactions of D-LSD with dopaminergic receptors. Brain Res 107: 303–315
Bockaërt J, Tassin JP, Thierry AM, Glowinski J, Prémont J (1977) Characteristics of DA and β-adrenergic sensitive adenylate cyclases in the frontal cerebral cortex of the rat. Comparative effects of neuroleptics on frontal cortex and striatal DA-sensitive adenylate cyclases. Brain Res 122: 71–86
Bowers MB, Salomonsson LA (1982) LSD: effect on monoamine metabolites in rat prefrontal cortex. Biochem Pharmacol 31: 4093–4096
Brozoski TJ, MC Brown R, Rosvold HE, Goldman PS (1979) Cognitive deficit caused by regional depletion of dopamine in prefrontal cortex of rhesus monkey. Science 205: 929–932
Brunello N, Barbaccia ML, Chuang DM, Costa E (1982) Down-regulation of β-adrenergic receptors following repeated injections of desmethylimipramine: permissive role of serotonergic axons. Neuropharmacology 21: 1145–1149
Bunney BS (1979) The electrophysiological pharmacology of midbrain dopaminergic systems. In: Horn AS, Korf J, Westerink BHC (eds) The neurobiology of dopamine. Academic Press, New York, pp 417–452
Bunney BS, Aghajanian GK (1976) DA and NA innervated cells in the rat prefrontal cortex. Pharmacological differentiation using microiontophoretic techniques. Life Sci 19: 1783–1792
Carter CJ, Pycock CJ (1980) Behavioural and biochemical effects of dopamine and noradrenaline depletion within the medial prefrontal cortex of the rat. Brain Res 192: 163–176
Cools AR, Van Dongen PAM, Janssen HJ, Megens AAPH (1978) Functional antagonism between DA and NA within the caudate nucleus of cats: a phenomenon of rhythmically changing susceptibility. Psychopharmacology 59: 231–242
Crosby G, Crane AM, Sokoloff L (1982) Local changes in cerebral glucose utilization during ketamine anesthesia. Anesthesiology 56: 437–443
Dickinson SL, Gadie B, Tulloch IF (1988) Alpha-1 and alpha-2 adrenoreceptor antagonists differentially influence locomotor and stereotyped behaviour induced by d-amphetamine and apomorphine in the rat. Psychopharmacology 96: 521–527
Ennis C, Kemp JD, Cox B (1981) Characterization of 5-hydroxytryptamine receptors that regulate dopamine release in the striatum. J Neurochem 36: 1505–1520
Fallon JH, Moore RY (1978) Catecholamine innervation of the basal forebrain. IV. Topography of the DA projection to the basal forebrain and neostriatum. J Comp Neurol 180: 545–580
Filion M, Tremblay L, Bédard PJ (1989) Excessive and unselective responses of medial pallidal neurons to both passive movement and striatal stimulation in monkeys with MPTP-induced parkinsonism. In: Neuronal mechanisms in disorders of movement. Libbey, London, pp 157–164
Freedman DX (1961) Effects of LSD25 on brain serotonin. J Pharmacol Exp Ther 134: 160–166
Freedman DX (1963) Psychotomimetic drugs and brain biogenic amines. Am J Psychiatry 119: 843–850
Freud S (1915) Das Unbewusste. Published in: Métapsychologie (1952) Gallimard, Paris, pp 91–161
Fuxe K, Hökfelt T, Johansson O, Jonsson G, Lidbrink P, Ljungdähl A (1974) The origin of DA nerve terminals in limbic and frontal cortex. Evidence for mesocortical neurons. Brain Res 82: 349–355

Galey D, Simon H, Le Moal M (1977) Behavioural effects of lesions of the DA A10 area of the rat. Brain Res 124: 83–97

Glowinski J, Tassin JP, Thierry AM (1984) The mesocorticoprefrontal dopaminergic neurons. Trends Neurosci 7: 415–418

Goodale MA, Milner AD, Jakobson LS, Carey DP (1991) A neurological dissociation between perceiving objects and grasping them. Nature 349: 154–156

Grace AA, Bunney BS (1983) Intracellular electrophysiology of nigral dopaminergic neurons. 3. Evidence for electrotonic coupling. Neuroscience 10: 333–348

Haigler HJ, Aghajanian GK (1977) Serotonin receptors in the brain. Fed Proc 36: 2159–2164

Hammer RP Jr, Herkenham M, Pert CB, Quirion R (1982) Correlation of regional brain metabolism with receptor localization during ketamine anaesthesia: combined autoradiographic 2(3H)deoxyglucose receptor binding technique. Proc Natl Acad Sci USA 79: 3067–3070

Hamon M (1984) Common neurochemical correlates to the action of hallucinogens. In: Jacobs BL (ed) Hallucinogens, neurochemical, behavioral, and clinical perspectives. Raven Press, New York, pp 143–169

Hervé D, Blanc G, Glowinski J, Tassin JP (1982) Reduction in DA utilization in the prefrontal cortex but not in the nucleus accumbens after selective destruction of noradrenergic fibers innervating the ventral tegmental area in the rat. Brain Res 237: 510–516

Hervé D, Trovero F, Blanc G, Thierry AM, Glowinski J, Tassin JP (1989) Non-dopaminergic prefrontocortical efferent fibers modulate D1 receptor deneveration supersensitivity in specific regions of the rat striatum. J Neurosci 9: 3699–3708

Hervé D, Trovero F, Blanc G, Vezina P, Glowinski J, Tassin JP (1990) Involvement of DA neurons in the regulation of β-adrenergic receptor sensitivity in rat prefrontal cortex. J Neurochem 54: 1864–1869

Hess EJ, Battaglia G, Norman AB, Ioro LC, Creese I (1987) Differential modification of striatal D1 dopamine receptors and effector moieties by N-ethoxycarbonyl-2-ethoxy-1,2-dihydroxyquinoline in vivo and in vitro. Mol Pharmacol 31: 50–57

Heym J, Trulson ME, Jacobs BL (1982) Raphe unit activity in freely moving cats: effects of phasic auditory and visual stimuli. Brain Res 232: 29–39

Hooper SL, Moulins M (1989) Switching of a neuron from one network to another by sensory-induced changes in membrane properties. Science 244: 1587–1589

Howard-Butcher S, Levine MS (1982) Behavioral and neurochemical effects of phencyclidine and its metabolites in the cat. Neurosci Abstr 103 (27.10)

Idapaan-Heikkila JE, Melsaac WM, Ho BT, Fritchie GE, Tansey LW (1969) Relation of pharmacology and behavioral effects of a hallucinogenic amphetamine to distribution in cat brain. Science 164: 1085–1087

Jacobs BL (1985) Overview of the activity of brain monoaminergic neurons across the sleep-wake cycle. In: Wauquier H, et al (eds) Sleep: neurotransmitters and neuromodulators. Raven Press, New York, pp 2–14

Johnson KM (1982) Acute and chronic phencyclidine administration: effects on the conversion of 3H-tryptophane to 3H-serotonin in rat forebrain. J Neural Transm 53: 179–186

Kelly PH, Seviour PW, Iversen SD (1975) Amphetamine and apomorphine responses in the rat following 6-OHDA lesion of the nucleus accumbens septi and corpus striatum. Brain Res 94: 507–523

Laplanche J, Pontalis JB (1968) Vocabulaire de la psychanalyse. Inconscient. PUF, Paris, pp 197–199

Lavielle S, Tassin JP, Thierry AM, Blanc G, Hervé D, Barthélémy C, Glowinski J (1979) Blockade by benzodiazepines of the selective high increase in DA turnover induced by stress in mesocortical DA neurons in the rat. Brain Res 168: 585–594

Lecanuet JP, Granier-Deferre J, Decasper AJ, Maugeais R, Andrieu AJ, Busnel MC (1987) Perception et discrimination foetales de stimuli langagier; mise en évidence à partir de la réactivité cardiaque; résultats préliminaires. CR Acad Sci Paris 305/III: 161–164

Le Moal M, Stinus L, Galey D (1976) Radio-frequency lesions of the ventral mesencephalic tegmentum: neurological and behavioural considerations. Exp Neurol 50: 521–535

Leonard BE (1973) Some effects of the hallucinogenic drug 2,5-dimethoxy-4-methylamphetamine on the metabolism of biogenic amines in the rat brain. Psychopharmacology 32: 33–49

Lewis DA, Campbell MJ, Foote SL, et al (1987) The distribution of tyrosine hydroxylase-immunoreactive fibers in primate neocortex is widespread but regionally specific. J Neurosci 7: 279–290

Magistretti PJ, Schorderet M (1985) Norepinephrine and histamine potentiate the increases in cyclic adenosine 3′:5′-monophosphate elicited by vasoactive intestinal polypeptide in mouse cerebral cortical slices: mediation by alpha 1-adrenergic and H1-histaminergic receptors. J Neurosci 2: 362–368

Martin LL, Smith DJ (1982) Ketamine inhibits serotonin synthesis and metabolism in vivo. Neuropharmacology 21: 119–125

Mason ST (1980) Noradrenaline and selective attention: a review of the model and the evidence. Life Sci 27: 617–631

Meibach RC, Glick SD, Cox R, Maayani S (1979) Localication of phencyclidine-induced changes in brain energy metabolism. Nature 282: 625–626

Meller E., Bohmaker K, Goldstein M, Friehoff AJ (1985) Inactivation of D1 and D2 dopamine receptors by N-ethoxy-carbonyl-2-ethoxy-1,2-dihydroquinoline in vivo: selective protection by neuroleptics. J Pharmacol Exp Ther 233: 656–662

Morrison JH, Molliver ME, Grzanna R, Coyle JT (1981) The intra-cortical trajectory of the coeruleo-cortical projection in the rat: a tangentially organized cortical afferent. Neuroscience 6: 139–158

Morrow AL, Creese I (1976) Characterization of alpha 1-adrenergic receptor subtypes in rat brain: a reevaluation of 3H-WB 4101 and 3H-Prazosin binding. Mol Pharmacol 29: 321–330

Nauta WJ, Fertag M (1979) The organization of the brain. Sci Am 241: 78–111

Persson SA (1978) Effects of LSD and BOL on the catecholamine synthesis and turnover in various brain regions. Psychopharmacology 59: 113–116

Peters DAV Tang S (1977) The effects of repeated D-lysergic acid diethylamide injections on catecholamine levels and tyrosine hydroxylase activity in rat brain regions. J Neurochem 28: 59–62

Piccolino M, Neyton J, Gerschenfeld HM (1984) Decrease of gap junction permeability induced by dopamine and cyclic 3′-5′-monophosphate in horizontal cells of turtle retina. J Neurosci 4: 2477–2488

Rasmussen K, Heym J, Jacobs BL (1984) Activity of 5-HT-containing neurons in nucleus centralis superior of freely moving cats. Exp Neurol 83: 302–317

Ravard S, Hervé D, Thiébot MH, Soubrié P, Tassin JP (1990) Anticonflict-like effect of a prefrontal DA lesion in rats: permissive role of NA neurons. Behav Pharmacol 1: 255–259

Romo R, Schultz W (1990) Dopamine neurons of the monkey midbrain: contingencies of responses to active touch during self-initiated arm movements. J Neurophys 63: 592–606

Rosecrans JA, Lovell RA, Freedman DX (1967) Effects of lysergic acid diethylamide on the metabolism of brain 5-hydroxytryptamine. Biochem Pharmacol 16: 2001–2021

Sachs O (1988) L'homme qui prenait sa femme pour un chapeau. Editions du Seuil, Paris, pp 251–272

Sawaguchi T, Goldman-Rakic PS (1991) D1 dopamine receptors in prefrontal cortex: involvement in working memory. Science 251: 947–950

Seeman P, Grigoriadis D (1987) Dopamine receptors in brain and periphery. Neurochem Int 10: 1–25

Servan-Schreiber D, Printz H, Cohen JD (1990) A network model of catecholamine effects: gain, signal-to-noise ratio, and behavior. Science 249: 892–895

Simon H, Le Moal M (1988) Mesencephalic dopaminergic neurons: role in the general economy of the brain. In: Kalivas PW, Nemeroff CB (eds) The mesocorticolimbic dopamine system. Ann NY Acad Sci 537: 235–261

Simon H, Le Moal M (1991) Mesocorticolimbic dopaminergic network: functional and regulatory roles. Physiol Rev: 155–233

Simon H, Scatton B, Le Moal M (1980) Dopaminergic A10 neurones are involved in cognitive functions. Nature 286: 150–151

Snyder SH, Reivich M (1966) Regional localization of lysergic acid diethylamide in monkey brain. Nature 209: 1093–1095

Sofia RD, Dixit BN, Barry III H (1971) The effect of delta 9 tetrahydrocannabinol on serotonin metabolism in the rat brain. Life Sci 10: 425–436

Sokoloff L, Reivich M, Kennedy C, Des Rosiers MH, Patlak CS, Pettigrew KD, Sakurada O, Shinoohara M (1977) Deoxyglucose method for the measurement of local cerebral glucose utilization: theory, procedure and normal value in the conscious and anaesthetized albino rat. J Neurochem 28: 897–916

Soubrié P (1986) Reconciling the role of central 5-HT neurons in human and animal behaviour. Brain Behav Sci 9: 319–335

Sporn JR, Harden TK, Wolfe BB, Molinoff PB (1976) β-Adrenergic receptor involvement in 6-OHDA induced supersensitivity in rat cerebral cortex. Science 194: 624–626

Steinfels GF, Heym J, Strecker RE, Jacobs BL (1983) Behavioural correlates of dopaminergic unit activity in freely moving cats. Brain Res 258: 217–228

Swanson LW, Hartman BK (1975) The central adrenergic system. An immunofluorescence study of the location of cell bodies and their efferent connections in the rat utilizing DA-β-hydroxylase as a marker. J Comp Neurol 163: 467–506

Taghzouti K, Simon H, Hervé D, Blanc G, Studler JM, Glowinski J, Le Moal M, Tassin JP (1988) Behavioural deficits induced by an electrolytic lesion of the rat ventral mesencephalic tegmentum are corrected by a superimposed lesion of the dorsal noradrenergic system. Brain Res 440: 172–176

Tassin JP (1989) Peut-on trouver un lien entre l'inconscient psychanalytique et les connaissances actuelles en neurobiologie? Neuro-Psy 4: 421–434

Tassin JP, Bockaërt J, Blanc G, Stinus L, Thierry AM, Lavielle S, Prémont J, Glowinski J (1978a) Topographical distribution of DA innervation and DA receptors of the anterior cerebral cortex of the rat. Brain Res 154: 241–251

Tassin JP, Lavielle S, Hervé D, Blanc G, Thierry AM, Alvarez C, Berger B, Glowinski J (1978b) Collateral sprouting and reduced activity of the rat mescortical dopaminergic neurons after selective destruction of the ascending noradrenergic bundles. Neuroscience 4: 1569–1582

Tassin JP, Stinus L, Simon H, Blanc G, Thierry AM, Le Moal M, Cardo B, Glowinski J (1978c) Relationship between the locomotor hyperactivity induced by A10 lesions and the destruction of fronto-cortical DA innervation in the rat. Brain Res 141: 267–281

Tassin JP, Simon H, Hervé D, Blanc G, Le Moal M, Glowinski J, Bockaért J (1982a) Non-dopaminergic fibers may regulate DA-sensitive adenylate cyclase in the prefrontal cortex and the nucleus accumbens. Nature 295: 696–698

Tassin JP, Simon H, Glowinski J, Bockaert J (1982b) Modulations of the sensitivity of DA receptors in the prefrontal cortex and the nucleus accumbens: relationship with locomotor activity. In: Collu R, Ducharme JR, Barbeau A, Tobis G (eds) Brain peptides and hormones. Raven Press, New York, pp 17–30

Tassin JP, Studler JM, Hervé D, Blanc G, Glowinski J (1986) Contribution of NA neurons to the regulation of DA (D1) receptor denervation supersensitivity in rat prefrontal cortex. J Neurochem 46: 243–248

Tassin JP, Hervé D, Vézina P, Trovero F, Blanc G, Glowinski J (1991) Relationships between mesocortical and mesolimbic dopamine neurons: Functional correlates of D1 receptor heteroregulation. In: Willner P, Scheel-Krüger J (eds) The mesolimbic dopamine system: from motivation to action. Wiley, Chichester, pp 175–196

Thierry AM, Blanc G, Sobel A, Stinus L, Glowinski J (1973) Dopaminergic terminals in the rat cortex. Science 182: 499–501

Tremblay L, Filion M (1989) Responses of pallidal neurons to striatal stimulation in intact waking monkeys. Brain Res 498: 1–16

Trovero F, Hervé D, Blanc G, Glowinski J, Tassin JP (1991) Different regulations of dopaminergic (D1) receptors and neurotensinergic binding sites in the rat prefrontal cortex. Neurosci Lett 127: 198–202

Trovero F, Blanc G, Hervé D, Vézina P, Glowinski J, Tassin JP (1992a) Contribution of an α_1-adrenergic receptor to the expression of the "ventral tegmental area syndrom". Neuroscience 47:69–76

Trovero F, Hervé D, Blanc G, Glowinski J, Tassin JP (1992b) A partial irreversible blockade of DA D1 receptors induces an hypersensitivity of DA-sensitive adenylate cyclase: involvement of alpha 1-adrenergic receptors. J Neurochem (in press)

Van Kammen DP, Antelman S (1984) Impaired noradrenergic transmission in schizophrenia? Life Sci 34: 1403–1413

Vézina P, Blanc G, Glowinski J, Tassin JP (1991) Opposed behavioural outputs of increased dopamine transmission in prefrontocortical and subcortical areas: a role for cortical D1 receptor. Eur J Neurosci 3: 1001–1007

Vickroy TW, Johnson KM (1982) Similar dopamine-releasing effects of phencyclidine and non-amphetamine stimulants in striatal slices. J Pharmacol Exp Ther 223: 669–674

Von Hungen K, Roberts S (1973) Adenylate cyclase receptors for adrenergic neurotransmitters in rat cerebral cortex. Eur J Biochem 36: 391–401

Von Hungen K, Roberts S, Hill DF (1975) Interactions between lysergic acid diethylamide and dopamine-sensitive adenylate cyclase systems in rat brain. Brain Res 94: 57–66

Wander TJ, Nelson A, Okazaki H, Richelson E (1987) Antagonism by neuroleptics of serotonin 5-HT1A and 5-HT2 receptors of normal human brain in vitro. Eur J Pharmacol: 279–282

Waterhouse BD, Hylan CM, Woodward DJ (1986) Interaction of serotonin with somatosensory cortical neuronal responses to afferent synaptic inputs and putative neurotransmitters. Brain Res Bull 17: 507–518

Weinberger DR (1986) Implications of normal brain development for the pathogenesis of schizophrenia. Arch Gen Psychiatry 44: 660–669

Woodward DJ, Moises HC, Waterhouse BD, Hoffer BJ, Freedman R (1979) Modulatory actions of norepinephrine in the central nervous system. Fed Proc 38: 2109–2116

Author's address: Dr. J. P. Tassin, Chaire de Neuropharmacologie, INSERM U.114, Collège de France, 11, Place Marcelin Berthelot, F-75005 Paris, France

J Neural Transm (1992) [Suppl] 36: 163–177

The anatomy of dopamine in monkey and human prefrontal cortex

P. S. Goldman-Rakic, M. S. Lidow, J. F. Smiley, and **M. S. Williams**

Section of Neurobiology, Yale School of Medicine, New Haven, CT, U.S.A.

Summary. This chapter reviews recent evidence establishing the comparable organization of dopamine afferents and dopaminergic receptors in the human and monkey prefrontal cortex. Light microscopy using a dopamine-specific antibody reveals that the dopamine innervation in the human prefrontal cortex exhibits a distinct bilaminar distribution with dense bands of fibers in the upper and deeper strata of the cortex, closely resembling the patterning of dopamine fibers in the monkey prefrontal cortex. Also, EM-immunohistochemistry has now revealed identical synaptic complexes both in human and monkey. In both species, dopamine axons from symmetric synapses predominantly on the spines of pyramidal cells. In many cases, the same spine is apposed by an asymmetric, putatively excitatory synapse. Finally, both in human and monkey prefrontal cortex, the dopamine D1-specific ligand, 3H-SCH23390, and the D2-specific ligand, H3-raclopride, label binding sites in laminar positions which match the location of the densest dopamine innervation. These results indicate that the organization of the cortical dopamine system is essentially the same in macaque monkey and human and that the nonhuman primate is a suitable animal model for analysis of dopamine function in prefrontal cortex.

Introduction

Dopamine deficiency or dysregulation has been suspected in the cognitive deficits associated with both Parkinson's disease and Schizophrenia as well as in a number of other diseases affecting cognition. Given the strong association between cognitive capacity and cortical function, it has long been evident that understanding the influence of dopamine on cortical structure and function would be a worthwhile goal. Although the earliest investigations of the monoamine innervation or content were conducted in rodents (for review, see Fallon and Loughlin, 1987), it has been known for some time that the differentiated primate cortex offers special opportunities for analysis of neurotransmitter interactions in the cortex (Brown and Goldman, 1976; Brown et al., 1977). Accordingly, over the past decade, we have sought to obtain more detailed knowledge of the dopamine distribution, its synaptic relationships and major postsynaptic targets in the primate cortex. Such information coupled with knowledge of local and long-tract connections, is essential for evaluating dopamine's precise contribution to cortical function.

Several aspects of the dopamine innervation in the primate cerebral cortex are now well recognized. Regional and/or laminar features have been documented by biochemical assay (Brown and Goldman, 1976; Brown et al., 1977), histo- and immunocytochemical methods (Levitt et al., 1984; Lewis et al., 1988) and radiographic studies of dopamine uptake (Berger et al., 1988). Furthermore, detailed anatomical data have also been supplemented by information on the functional consequences of perterbations in the dopamine innervation of the cortex. Experimentally induced dopamine depletion in the prefrontal cortex results in memory loss comparable in extent to that produced by excision of the principal sulcus (Brozoski et al., 1979). Additionally, newly synthesized compounds with specific agonist and antagonist properties (e.g., Hyttel, 1982; Iorio et al., 1983) have opened up opportunities for analysis of neurotransmitter receptors in both anatomical (Goldman-Rakic et al., 1990) and behavioral (Sawaguchi and Goldman-Rakic, 1991) studies whereas antibodies raised specifically against dopamine and dopaminoceptive neurons have enhanced our ability to localize dopamine in axon terminals and in synaptic complexes of the rhesus monkey cortex (Goldman-Rakic et al., 1987).

One lingering question in the detailed study of these issues in the nonhuman primate is the degree to which they reflect the status of dopamine afferents, dopamine synapses and dopamine receptors in human cortex. Although the monkey is ideal for relating the functional properties of prefrontal cortex to the prefrontal cortical circuitry which has been worked out in some detail (Goldman-Rakic, 1987), the human cortex is larger and may differ in a species typical manner from its simian relatives. This chapter reviews recent evidence from my laboratory on the distribution of dopamine-containing fibers, their mode of synaptic termination, their synaptic targets and the laminar disposition of dopaminergic receptors in the prefrontal cortex of nonhuman primates and compares these findings with recent results obtained with basically identical methods and protocols in human neocortex. The prefrontal cortex has a well documented role in cognitive functions (Milner, 1964; Fuster, 1980; Goldman-Rakic, 1987), and thus it is a favorable structure in which to examine the influence that dopamine transmission may have on cognitive function.

Distribution of dopamine afferents in the prefrontal cortex: light microscopy

Nonhuman primate

In rhesus monkey, the dopamine afferents in prefrontal cortex exhibit a characteristic bilaminar distribution of high concentration in layers I, II &

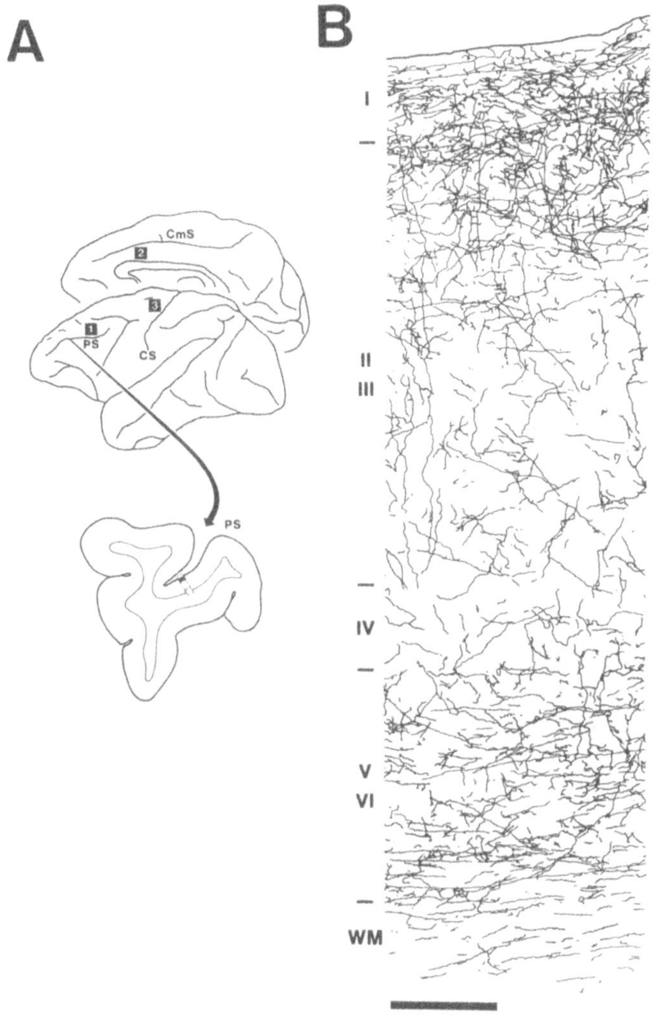

Fig. 1. Camera lucida drawing of dopamine immunopositive fibers in the principal sulcus of a rhesus monkey showing high density of fiber plexes in layers I and II/III. Many of those in the upper part of layer I run parallel to the pia for long distances. Fiber density in lower parts of layer III and in layer IV are notably thinner than in the surrounding laminae. In the deeper layers V and VI, dopaminergic fiber systems are more dense with a mixture of tangential, oblique and vertically directed processes, consistent with the pattern of basilar dendrites belonging to the projection neurons in these layers. Bar = 200 µm. [Drawing by M. Williams from Goldman-Rakic et al. (1989)]

upper III and also in layers V and VI (Fig. 1). The intermediate layers of the cortex, including the widest part of layer III contain an ample dopamine innervation but this is still considerably less than in the adjacent strata. The bilaminar pattern has been documented with a dopamine-specific monoclonal antibody (Goldman-Rakic et al., 1989), with antibodies raised against tyrosine-hydroxylase (TH) (Lewis et al., 1988) and with the method of autoradiographic uptake of [^{3}H]-dopamine (Berger et al., 1986, 1988). Thus, three independent methods have obtained the same basic result of a bilaminar pattern of fibers basically throughout the monkey prefrontal cortex.

Human

Since interest in the dopamine innervation of cortex stems in major part from its relevance to human mentation, a major question has been whether the cortical dopamine innervation is as substantial in humans as in monkeys and whether it follows the same bilaminar profile of distribution. Recently, we have had the opportunity to apply the DA-specific antibody at the light microscopic level to biopsy tissue obtained from neocortical areas of the frontal lobe and temporal pole during the course of surgical resection for relief of intractable seizures (Smiley et al., 1992). The tissue was obtained through the kind cooperation of Dr. Dennis Spencer, Chief, Section of Neurosurgery, Yale University. The tissues studied are not in the epileptogenic focus and appear in Nissl stained sections to be normal in cytoarchitectionic appearance. Moreover, the pattern of dopamine innervation observed in the human prefrontal and temporal cortex was remarkably similar to that seen in the rhesus monkey (Fig. 2). Again, the highest concentration of labeled fibers was in the upper layers I, II and upper III; and also in layers V and VI. As in the monkey, the intermediate layers had a noticeably weaker dopamine innervation (Fig. 2). Thus, at the light microscopic level, the dopamine innervation in humans appears remarkably similar to that observed in nonhuman primates.

Dopamine axon terminals: electron microscopy

Nonhuman primate

To understand how dopamine influences cortical function, it is necessary to know the precise target structures which receive dopamine afferents. Thus, our laboratory has been devoting itself to analysis of macaque monkey prefrontal cortex at the ultrastructural level. This examination has revealed that pyramidal neurons are the principle target of the dopamine innervation. Dopamine terminals in the prefrontal cortex are of the symmetric type and they are largely found on the dendrites and spines of the pyramidal cells (Goldman-Rakic et al., 1989). In rodents. DA synapses are also symmetric but fewer of them appear to be made on spines relative to dendrites and soma (Van Eden et al., 1987; Segula et al., 1988). Nevertheless, data in rats and monkeys allow the inference that dopamine release in the cortex can have a direct and powerful influence on cortical neuronal activity. An unexpected and potentially important result from our study in monkeys is

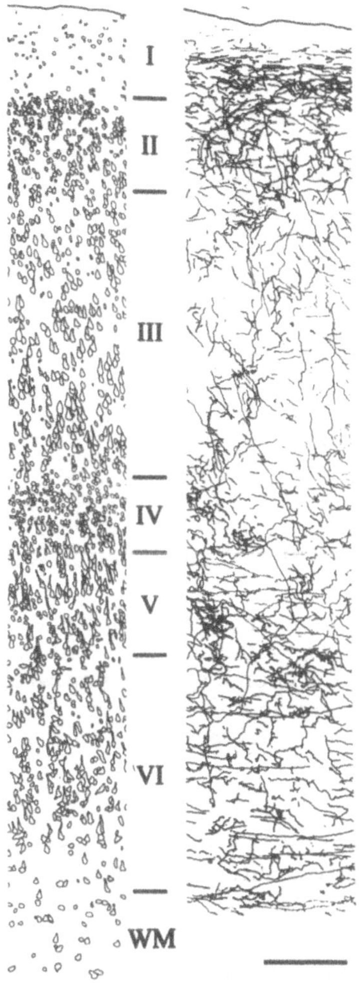

Fig. 2. Camera lucida drawing of cell density (left) and dopamine immunopositive fibers (right) in the cortex of a human biopsy specimen. As in the monkey, dopamine fiber density is highest in the upper and deeper strata and less dense in the intermediate layers, including all of layer III. Bar = 200 μm. [Drawing by M. Williams from Smiley et al. (1992)

shown in Fig. 3. We noticed that the spines of pyramidal neurons were invariably the targets of both DA afferents and another unspecified asymmetric (presumed excitatory) synapse (Goldman-Rakic et al., 1989). Such synaptic "triads" were found in prefrontal, premotor and motor cortex suggesting that the anatomical arrangement may be widespread and common to many cortical areas. Synaptic triads in which the dendritic spines of medium spiny neurons receive both a TH-positive bouton and an asymmetric synaptic specialization are found in the rodent neostriatum (Freund et al., 1984) and the rat prefrontal cortex also may contain synaptic complexes with DA-positive profiles, similar to that we observed in the monkey

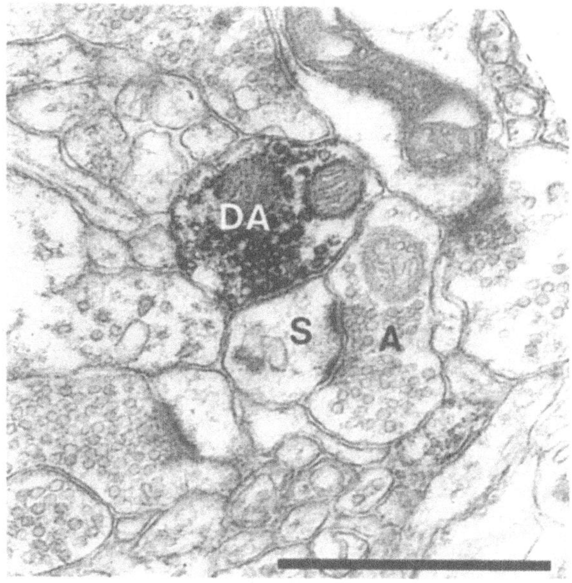

Fig. 3. Electron micrograph of a dopamine-immunostained profile taken from layer II of the monkey prefrontal cortex, illustrating a characteristic synaptic triad composed of a da-positive symmetric synapse (*DA*) on a spine (*S*) on which an asymmetric synapse (*A*) of unidentified origin is also present. Bar = 1 μm. [From Goldman-Rakic et al. (1989)]

(Verney et al., 1990). Thus, in both the cortex and striatum, the DA terminal endings on spines are strategically placed within the local circuitry to alter local spine responses to excitatory inputs, and through these, to make a contribution to the overall excitability of projection neurons. Presumably both facilitatory and inhibitory effects of dopamine might be predicted. In the rat, DA release in the cortex can be occasioned by VTA stimulation or can be simulated by local iontophoretic application of dopamine; these manipulations in the rat both reduce spontaneous activity and dampen the excitatory responses produced by thalamic stimulation or by tail pinch (Ferron et al., 1984).

It is also tempting to relate the triadic synaptic arrangement in the cerebral cortex to the finding of Halpain and Greengard on the antagonistic effects of dopaminergic D1 and glutamate NMDA receptor stimulation on cyclic-AMP mediated second messengers of target neurons (Halpain et al., 1990). It may be that the triad is an anatomical substrate for D1/glutamate interaction in the cortex.

Human

Again, we were interested in whether such arrangements could be observed in human cortex. However, ultrastructural studies in human brain are technically difficult because the tissue is not optimally fixed by immersion. However, as mentioned, biopsy tissue obtained during tumor excision or neurosurgical treatment for intractable epilepsy appears normal on histological sections. Using the dopamine-specific antibody, we have been able to visualize dopamine profiles at the ultrastructural level in this tissue. To our knowledge this is the first documentation of dopamine synapses in human cortex (Fig. 4; Smiley et al., 1992). Moreover, we found the dopamine synapses to be distributed in a manner similar to that which we documented in monkeys — namely the synapses were of the symmetric type and their predominant targets were foremost the spines and to a somewhat lesser extent the dendrites of pyramidal neurons. As in the monkey, non-symmetric synapses were almost never observed contacting cell bodies. The principal difference which could be observed between human and monkey dopamine synapses were that the former were larger, and a small proportion of them were asymmetric. Of particular interest was that among the more than 60% that formed axospinous synapses, some of them entered into triadic complexes with asymmetric synapses on the same spine, much as we observed in the monkey. We believe that the triadic arrangement is important for direct modulation by dopamine of "sensory" integration across multiple inputs in the cortex (Goldman-Rakic et al., 1989).

Distribution of the dopaminergic D1 and D2 receptors: receptor autoradiography

Nonhuman primates

The areal and laminar distribution of the major dopaminergic, neurotransmitter receptors have been examined throughout the various cytoarchitectonic regions of the prefrontal cortex with the dopamine selective ligands, [^{3}H]SCH-23390, and [^{3}H]raclopride to label the dopaminergic D-1 and D-2 dopamine receptor subtypes, respectively (Goldman-Rakic et al., 1990; Lidow et al., 1991). Quantitative densitometry revealed that each dopaminergic site is distributed in a distinctive laminar-specific pattern that is remarkably similar from area to area within the prefrontal cortical domain. The D-1 dopamine receptor is present in highest relative concentration in superficial layers I, II and IIIa while the D-2 receptor is distinguished by relatively high concentrations in deep layer V compared to all other layers (Fig. 5). The localization of D-1 and D-2 sites together thus corresponds reasonably well to the bilaminar distribution of the DA innervation described above. It is important to note that the density of the two major subtypes of dopamine receptor differed dramatically — with the D1 sites having an approximately 10× higher concentration than the D2 sites. This widespread and more dense distribution of D-1 receptors relative to D-2 receptors in the primate prefrontal cortex is relevant to our recent finding that intracerebral injection of the dopamine D1 antagonist,

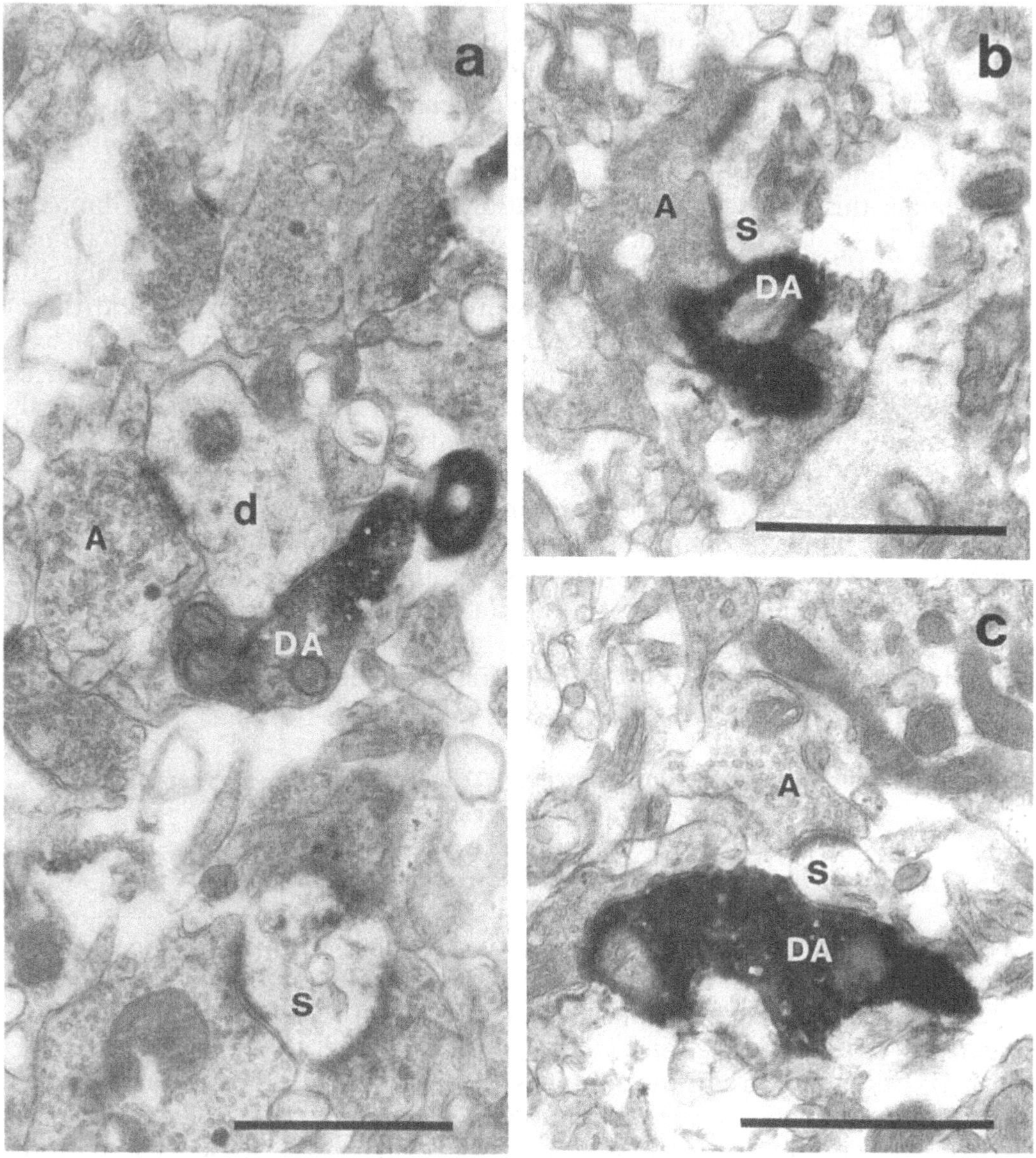

Fig. 4. Electron micrograph of a dopamine-immunostained profile taken from the temporal cortex of a human biopsy specimen. The photomicrograph illustrate a synaptic triad similar to that consistently found in the monkey cortex. Again the triad is composed of a dopamine-positive symmetric synapse (*DA*) on a spine (*2*) on which an asymmetric synapse (*A*) of unidentified origin is also present. Bar = 1 μm. [From Smiley et al. (1992)

SCH23390 into area 46 of rhesus monkeys produced an increased latency and decreased accuracy of response in delayed-response performance whereas D-2 antagonists were without effect (Fig. 6, Sawaguchi and Goldman-Rakic, 1991). At present it is not possible to determine whether the various receptors are located presynaptically on the afferent terminals or postsynaptically on cortical neurons. Future studies using *in situ* hybridization may answer this question.

That D-2 binding sites are present in layer V throughout the prefrontal cortex (and in other areas) bears emphasis because the very existence of D-2 receptors in any cortical area, including prefrontal cortex, had been in considerable doubt (see Fallon and Loughlin, 1987 for recent review). However, in our receptor autoradiographic study in primates, we consistently found a low density of D-2 receptors in prefrontal cortex (Goldman-Rakic et al., 1990) and similarly low concentrations were observed in recent homogenate binding studies with the same D-2-specific antagonist (Lidow et al., 1989d). These findings concur with an earlier study of rodent cortex using the D-2-specific ligand, [^{3}H]iodosulpiride (Martres et al., 1985). Together with the results in the monkey, it appears that the difficulty in detecting D-2 receptors in the cortex may simply be due to their low concentration or perhaps to the use of nonspecific ligands. However, in spite of the low concentrations of the D-2 receptors, their strict laminar localization is of some interest because cells in layer V are specifically associated with motor outflow through the major cortico-striatal and cortico-tectal projections and to a lesser extent with cortico-cortical fibers (Schwartz and Goldman-Rakic, 1984).

Human

To turn to the distribution of D1 and D2 dopaminergic receptor subtypes in human prefrontal cortex, again we find that there is good correspondence with the results described for rhesus monkey (Fig. 5). As in monkey, we used tritiated raclopride to label D2 and SCH23390 to label D1 receptors. The D1 receptor was concentrated in the superficial strata and in deep layers V and VI whereas the D2 receptor labeling was concentrated in a thin line throughout layer V. Correspondence in the distribution of these receptors in the human and monkey frontal lobe is striking and all the more so since some differences were found in the laminar distributions of these receptors in other cortical regions. For example, in the perirhinal, motor and visual areas, the distribution of the D1 receptor was notably different in human and monkey (Lidow and Goldman-Rakic, unpublished observations).

D-1 bearing neurons in prefrontal cortex: DARPP-32 labeling

The dopamine and cyclic AMP-regulated phosphoprotein of molecular weight 32K (DARPP-32), which has been found to be enriched in neurons that bear D1 receptors (Walaas et al., 1983) has recently been examined in the cerebral cortex both in rat (Ouimet et al., 1984) and monkey (Ouimet et al., 1992, in press; Berger et al., 1990). In both species, labeled cells were

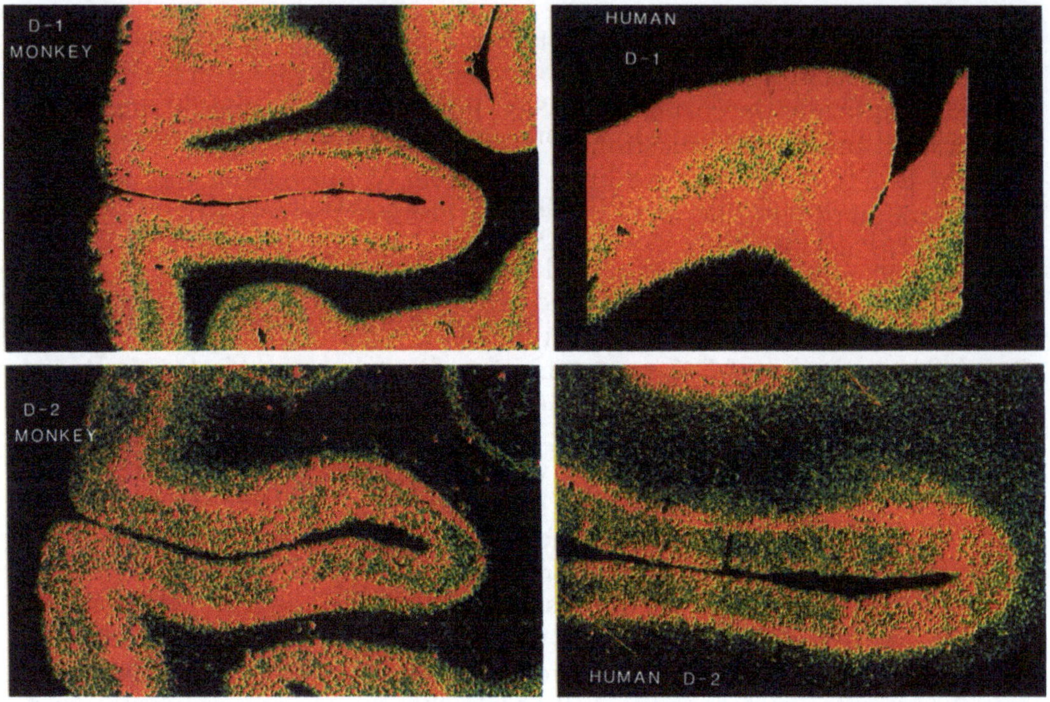
D-1
MONKEY
HUMAN
D-1
D-2
MONKEY
HUMAN D-2

Fig. 5

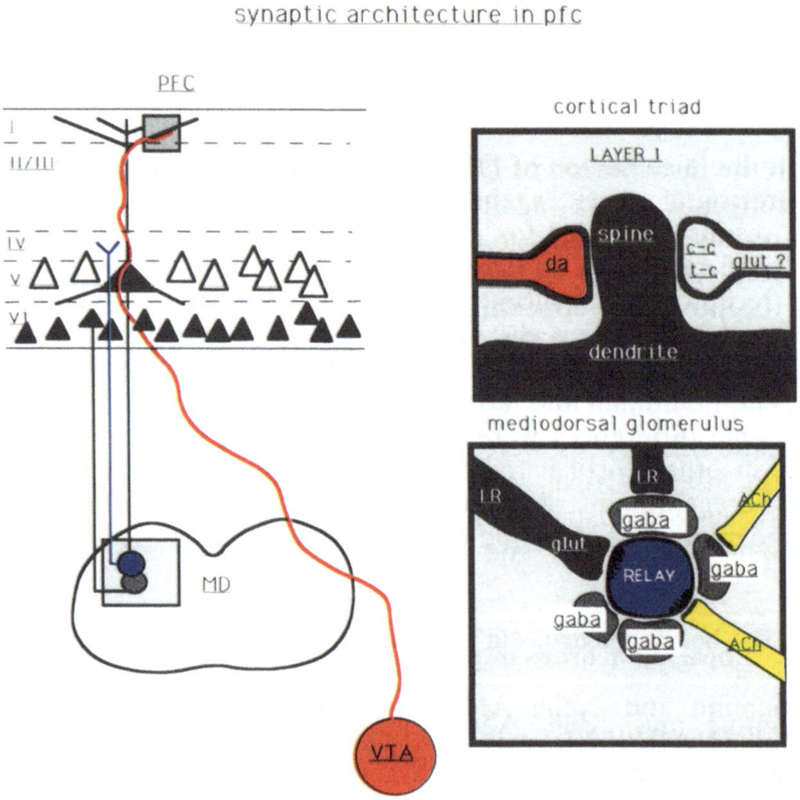
synaptic architecture in pfc
PFC
I
II/III
IV
V
VI
MD
VTA
cortical triad
LAYER I
spine
da
c-c
t-c
glut ?
dendrite
mediodorsal glomerulus
LR
LR
ACh
gaba
glut
RELAY
gaba
gaba
gaba
ACh

Fig. 8

observed to be preferentially located in layers II and V/VI, thus observing a bilaminar distribution, corresponding in part to the dual laminar pattern of dopamine innervation and D-1 receptor localization. It is of interest that neurons in layer VI are known to project to the thalamus and recent evidence indicates that a subset of neurons in layer V likewise project to the thalamus (Fig. 7, Giguere and Goldman-Rakic, 1988). The apical dendrites of the DARPP-32 positive neurons ascend to and branch in the upper layers where their spines are the targets of thousands of incoming afferent inputs. It seems a reasonable hypothesis that the DARPP-32 positive neurons in the cortex are a primary target of the cortical dopamine innervation. Whether D2 receptors are located on the same or a different subset of cortical neurons is not known.

Complimenting the evidence for a dual source of projections to the thalamus from layers V and VI (Giguere and Goldman-Rakic, 1988) is the recent finding that their are also two types of cortical synapses in the mediodorsal nucleus of the thalamus (Schwartz et al., 1991). In this study, the prefrontal cortex was injected with tritiated amino acids and EM — autoradiography revealed two types of labeled synapses in the MD — a large profile that contacts the primary dendrite of the thalamo-cortical relay neuron which is a central element in a thalamic grouping called a glomerulus and a smaller terminal bouton that forms an asymmetric synapse on presynaptic (probably GABA-containing) dendrites that lay within the glomeruli (Fig. 8). Thus, the cortico-thalamic projections of the prefrontal

Fig. 5. Photographs in pseudocolor of autoradiograms from rhesus monkey (left) and human (right) displaying the distribution of 3H-SCH33290 (D1) and 3H-raclopride binding sites in prefrontal cortex of each species. The similarity is quite striking; D1 sites are mainly located in the extreme layers, corresponding to the pattern of innervation shown in Figs. 1 and 2; the most concentrated D2 sites form a narrow band in upper layer V in both rhesus monkey and in human

Fig. 8. Diagram of the DARPP-32 positive neurons in the cortex and their hypothetical relationships with dopamine input (red) from the ventral tegmental area (VTA) and neuronal elements in the mediodorsal thalamic nucleus (MD). *Upper right hand panel*: cartoon drawing of a common synaptic triadic arrangement found in both human and monkey prefrontal cortex: a dopamine symmetric synapse and an unidentified asymmetric synapse are apposed on the same spine; the diagram depicts the *hypothesis* that a layer V, DARPP-32 immuno-positive neuron (presumably bearing D1 receptors) is a primary target of the dopamine axospinous synapse. *Lower right hand panel*: illustration of the glomerular arrangement of GABA-containing presynaptic dendrites which form symmetric (putatively inhibitory) synapses on the principal relay neuron in the mediodorsal nucleus. EM studies reveal that the corticothalamic fibers can influence the principal relay neuron directly via excitatory synaptic contacts and indirectly via inhibitory interneurons (Schwartz et al., 1991). If the cortico-thalamic projection neurons are DARPP-32 positive, as the immunohistochemical evidence suggests, then the same cortical neurons that project upon the thalamus (as well as other subcortical structures) may be a major target of the dopamine innervation

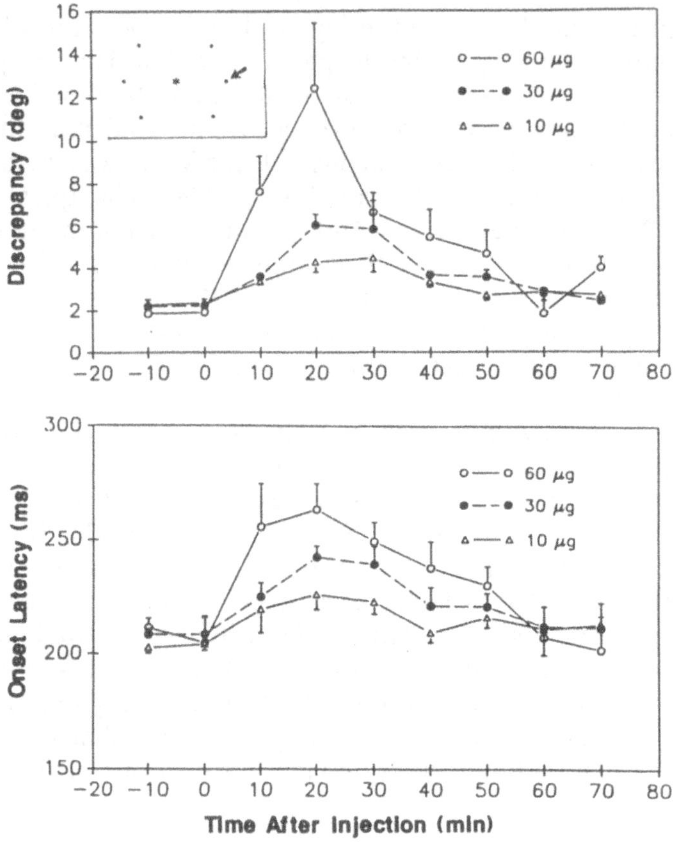

Fig. 6. Injection of the D1 antagonist, SCH23390, into prefrontal cortex of rhesus monkeys induced a selective deficit in their performance of memory-guided saccades. The deficit consisted of a decrease in the accuracy of response (top) as well as a slowing of the latency to respond (bottom) on each trial. The distribution of SCH23390 binding sites in the prefrontal cortex is shown in Fig. 5. [Data from Sawaguchi and Goldman-Rakic (1991)]

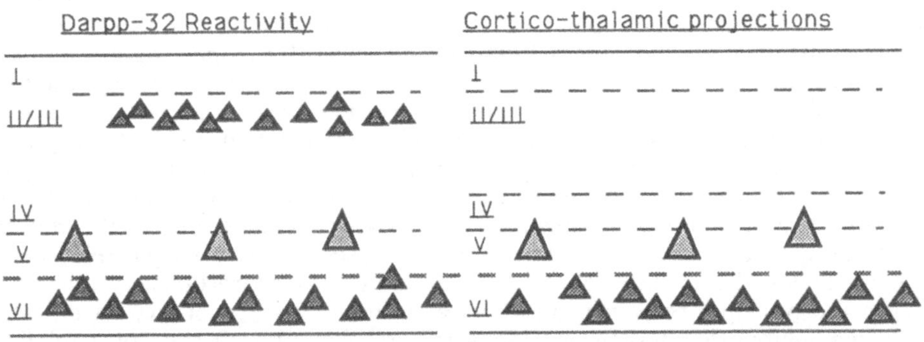

Fig. 7. Diagrams of Darpp-32 labeled cells in the primate cortex (from Berger et al., 1991); and of the corticothalamic projection neurons from the same area identified by retrograde tracing from the thalamus (From Giguere and Goldman-Rakic, unpublished)

cortex can exert both direct excitatory influence and an indirect inhibitory influence on the very principal relay cell that projects back to the cortex. This ultrastructural study thus provides evidence of a reciprocal connection between the thalamo-cortical and cortico-thalamic connections in primate prefrontal cortex and also is suggestive of a mechanism for gating thalamic input.

Conclusions and speculations relating to dopamine function

Anatomical and receptor binding data is converging to provide a picture of the cortical circuits influenced by dopamine in the prefrontal cortex. Although the picture is still lacking in important details, we can begin to appreciate that dopamine's influence may be highly specific. If we regard the cortical innervation specifically, we would be inclined to suggest that the dopamine innervation is targeted to influence non-thalamic circuitry because the lightest concentration of dopamine fibers and synapses is in layer IV. On the other hand, this may be totally misleading, particularly if we consider that dopamine fibers may be influencing predominantly those neurons that project to the thalamus and, in addition, the basal ganglia and tectum (layer V cells). The effect of cortical dopamine may thus be to modulate and/or adjust the gain on a number of output pathways, possibly enhancing, possibly dampening, possibly disrupting information flow through these centers. As the basal ganglia-thalamo-cortical loop system ultimately controls motoric action, dopamine's role in motor control perhaps can be extended to the memory-guided behaviors of the prefrontal cortex in a powerful and direct way.

Our findings on the higher density of D1 receptors in prefrontal cortex draw attention to the potential functional significance of these receptors for the cognitive deficits observed in many diseases affecting the cortex. If we suppose that there is an optimal level of dopamine stimulation necessary for normal cognitive processing, we could envision that both loss of cortical dopamine (e.g. in Parkinson's disease) or increased release (e.g. amphetamine psychosis) may be detrimental to the thought process. Our finding that dopamine and putative glutamate profiles are apposed to the membrane surface of the same dendritic spine gives rise to hypotheses concerning the balance of glutamate and dopamine receptors. Finally, if D1 receptors are important targets of atypical neuroleptics, and there is evidence that clozapine has high affinity for D1 dopaminergic receptors as well as for other receptors, the corticothalamic pathway may turn out to be a specific site of dysfunction in schizophrenia and a specific target for therapeutic measures.

References

Berger B, Trottier S, Gaspar P, Verney C, Alvarez C (1986) Major dopamine innervation of the cortical motor areas in the cynomolgus monkey. Neurosci Lett 72: 121–127

Berger B, Trottier S, Verney C, Gaspar P, Alvarez C (1988) Regional and laminar distribution of dopamine and serotonin innervation in the cynomolgus cerebral cortex. Major differences of dopamine input in the granular and agranular cortices. A radioautographic study. J Comp Neurol 273: 99–119

Berger B, Febvret A, Greengard P, Goldman-Rakic PS (1990) DARPP-32, a phosphoprotein enriched in dopaminoceptive neurons bearing dopamine D1 receptors: distribution in the cerebral cortex of the newborn and adult rhesus monkey. J Comp Neurol 299: 327–348

Brown RM, Goldman PS (1976) Distribution and development of catecholamines in primate neocortex. Soc Neurosci Abstr 2: 469

Brown RM, Crane AM, Goldman-Rakic PS (1977) Catecholamines in neocortex of rhesus monkeys: regional distribution and ontogenetic development. Brain Res 124: 576–580

Brown RM, Crane AM, Goldman PS (1979) Regional distribution of monoamines in the cerebral cortex and subcortical structures of the rhesus monkey: concentrations and in vivo synthesis rates. Brain Res 168: 133–150

Brozoski T, Brown RM, Rosvold HE, Goldman PS (1979) Cognitive deficit caused by depletion of dopamine in prefrontal cortex of rhesus monkey. Science 205: 929–932

Fallon JH, Loughlin SE (1987) Monoamine innervation of cerebral cortex and a theory of the role of monoamines in cerebral cortex and basal ganglia. In: Jones EG, Peters A (eds) Cerebral cortex, vol 6. Plenum, New York, pp 41–127

Ferron A, Thierry AM, Le Douarin C, Glowinski J (1984) Inhibitory influence of the mesocortical dopaminergic system on spontaneous activity or excitatory response induced from the thalamic mediodorsal nucleus in the rat medial prefrontal cortex. Brain Res 302: 257–265

Freund TF, Powell JF, Smith AD (1984) Tyrosine hydroxylase-immunoreactive boutons in synaptic contact with identified striatonigral neurons, with particular reference to dendritic spines. Neuroscience 13: 1189–1215

Fuster JM (1980) The prefrontal cortex. Raven Press, New York

Giguere M, Goldman-Rakic RS (1988) Mediodorsal nucleus: areal, laminar and tangential distribution of afferents and efferents in the frontal lobe of rhesus monkeys. J Comp Neurol 277: 195–213

Goldman-Rakic PS (1987) Circuitry of primate prefrontal cortex and regulation of behavior by representational memory. In: Plum F (ed) Handbook of physiology vol 5. The nervous system, higher functions of the brain ch 9. Am Physiol Soc, Bethesda MD, pp 373–417

Goldman-Rakic PS, Leranth C, Williams MS, Mons N, Geffard M (1989) Dopamine synaptic complex with pyramidal neurons in primate cerebral cortex. Proc Natl Acad Sci 86: 9015–9019

Goldman-Rakic PS, Lidow MS, Gallager DW (1990) Overlap of dopaminergic, adrenergic, and serotoninergic receptors and complementarity of their subtypes in primate prefrontal cortex. J Neurosci 10: 2125–2138

Halpain S, Girault JA, Zorn S, Greengard P (1990) Activation of NMDA receptor induces dephosphorylation of DARPP-32 in rat striatal slices. Nature 343: 369–372

Hyttel J (1982) Citalopram: basic and clinical studies. Prog Neuropharmacol Biol Psychiatry 6: 257–336

Iorio IC, Barnett A, Leitz FH, Houser VP, Corduba CA (1983) SCH23390, a potential benzodiazepine antipsychotic with unique interactions on dopaminergic system. J Pharmacol Exp Ther 226: 462–468

Levitt P, Rakic P, Goldman-Rakic PS (1984) Region-specific distribution of catecholamine afferents in primate cerebral cortex: a fluorescence histochemical analysis. J Comp Neurol 227: 23–36
Lewis DA, Foote SL, Goldstein M, Morrison JH (1988) The dopaminergic innervation of monkey prefrontal cortex: a tyrosine hydroxylase immunohistochemical study. Brain Res 449: 225–243
Lidow MS, Goldman-Rakic PS, Rakic P, Innis R (1989) Dopamine D2 receptors in the cerebral cortex: distribution and pharmacological characterization with [^{3}H]raclopride. Proc Natl Acad Sci 86: 6412–6416
Lidow MS, Goldman-Rakic PS, Gallager DW, Rakic P (1991) Distribution of dopaminergic receptors in the primate cerebral cortex: quantitative autoradiographic analysis using [^{3}H]raclopride, [^{3}H]spiperone and [^{3}H]SCH23390. Neuroscience 40: 657–671
Martres MP, Sales N, Bouthenet ML, Schwartz JC (1985) Localization and pharmacological characterization of D-2 dopamine receptors in rat cerebral neocortex and cerebellum using [^{125}I]iodosulpiride. Eur J Pharmacol 118: 211–219
Milner B (1964) Some effects of frontal lobectomy in man. In: Warren JM, Akert K (eds) The frontal granular cortex and behavior. McGraw-Hill, New York, pp 313–334
Ouimet CC, Miller PE, Hemmings HC Jr, Walaas SI, Greengard P (1984) DARPP-32, a dopamine- and adenosine 3′:5′-monophosphate-regulated phosphoprotein enriched in dopamine-innervated brain regions. III. Immunocytochemical localization. J Neurosci 4: 111–124
Sawaguchi T, Goldman-Rakic PS (1991) D1 dopamine receptors in prefrontal cortex: involvement in working memory. Science 251: 947–950
Schwartz ML, Goldman-Rakic PS (1984) Callosal and intrahemispheric connectivity of the prefrontal association cortex in rhesus monkey: relation between intraparietal and principal sulcal cortex. J Comp Neurol 226: 403–420
Schwartz ML, Dekker JJ, Goldman-Rakic PS (1991) Dual mode of corticothalamic synaptic termination in the mediodorsal nucleus of the rhesus monkey. J Comp Neurol 309: 289–304
Seguela P, Watkins KC, Descarries L (1988) Ultrastructural features of dopamine axon terminals in the anteromedial and suprarhinal cortex of adult rat. Brain Res 442: 11–22
Smiley JF, Williams SM, Szigeti K, Goldman-Rakic PS (1992) Light and electron microscopic characterization of dopamine-immunoreactive processes in human cerebral cortex. Comp Neurol (in press)
Van Eden CG, Hoorneman, EMD, Buijs RM, Matthissen MAH, Geffard M, Uylings HBM (1987) Immunocytochemical localization of dopamine in the prefrontal cortex of the rat at the light and electron microscopical level. Neuroscience 22: 849–862
Verney C, Alvarez C, Geffard M, Berger B (1990) Ultrastructural double-labelling study of dopamine terminals and GABA-containing neurons in rat anteromedial cerebral cortex. Eur J Neurosci 2: 960–972
Walaas SI, Aswad DW, Greengard P (1983) AMP-regulated phosphoprotein enriched in dopamine-innervated brain regions. Nature 301: 69–71

Author's address: Dr. P. S. Goldman-Rakic, Section of Neurobiology, Yale School of Medicine, 333 Cedar Street, New Haven, CT 06510, U.S.A.

J Neural Transm (1992) [Suppl] 36: 179–200

The catecholaminergic innervation of primate prefrontal cortex

D. A. Lewis

Departments of Psychiatry and Behavioral Neuroscience, University of Pittsburgh, Pittsburgh, PA, U.S.A.

Summary. This paper reviews recent studies indicating that the marked expansion and differentiation of the prefrontal cortex in primates is associated with an increase in the complexity of both the regional density and laminar distribution of catecholaminergic afferents. The innervation patterns of these systems in monkey prefrontal cortex appear to accurately predict those in human prefrontal cortex, suggesting that studies in non-human primates may be reasonably used to generate hypotheses about the nature of involvement of these systems in disorders such as schizophrenia. In addition, the distinctive developmental pattern of the dopaminergic innervation of primate prefrontal cortex and the possibility of an intrinsic catecholaminergic innervation of primate prefrontal cortex may reveal new avenues of investigation into the roles of prefrontal catecholamines in both normal and pathological states.

Introduction

In comparison to other species, one of the most striking features of the primate brain is the marked expansion and differentiation of the prefrontal cortical regions. These neocortical regions are located anterior to the motor and premotor areas of the frontal lobe on its lateral, medial and orbital surfaces (Fig. 1). The substantially greater relative size of the prefrontal cortex in the human brain has led to the inference that these cortical regions are the neural substrate for some types of high level cognitive processing. Although the precise functions of the prefrontal cortex remain enigmatic, many experimental findings support the view that the prefrontal regions are important mediators of certain complex behaviors, such as those involving temporal integration (Fuster, 1989) and working memory (Goldman-Rakic, 1987a).

These observations suggest that the prefrontal regions may be involved in the pathophysiology of human neuropsychiatric disorders that disrupt higher cognitive functions. In the case of schizophrenia, a variety of lines of evidence suggest that certain prefrontal regions may be a locus of dysfunc-

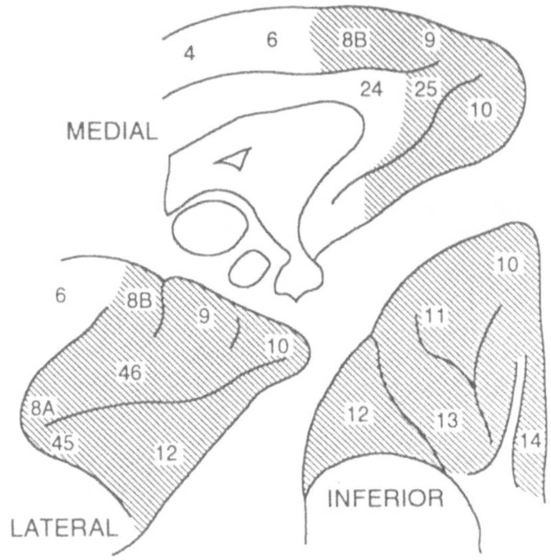

Fig. 1. Map of the cytoarchitectonic areas of the macaque prefrontal cortex, adapted from Walker (1940)

tion or structural pathology in this disorder. For example, functional imaging studies have demonstrated that schizophrenic patients perform poorly, and fail to activate the dorsolateral prefrontal cortex while engaged in a task, the Wisconsin Card Sort Test, which depends upon the functional integrity of this prefrontal region (Weinberger et al., 1986). This dysfunction has been suggested to be related to abnormalities in monoaminergic systems (Weinberger et al., 1988). In addition, based primarily upon pharmacological data, the catecholamines, norepinephrine and especially dopamine, have also been implicated in the pathophysiology of schizophrenia (see Wyatt et al., 1985 for review).

Although many aspects of the catecholaminergic innervation of rodent prefrontal cortex have been extensively investigated, the greater size and complexity of primate prefrontal cortex has been found to be associated with a number of differences in the catecholaminergic innervation of these regions. In this paper, we review recent investigations regarding the catecholaminergic innervation of monkey and human prefrontal cortex. The findings of these studies may have important implications for the generation of novel hypotheses regarding the role of prefrontal cortex and catecholamines in the pathophysiology of schizophrenia.

Selective identification of dopaminergic and noradrenergic axons in primate neocortex

The prefrontal cortex receives axonal projections from the noradrenergic neurons of the locus coeruleus and from the dopaminergic neurons of the ventral mesencephalon. Both biochemical (Bjorklund et al., 1978; Brown et al., 1979) and catecholamine histofluorescence studies (Levitt et al., 1984) have confirmed the presence of these catecholamines in monkey prefrontal cortex. In order to identify the regional and laminar patterns of innervation of monkey prefrontal cortex by catecholaminergic axons, we used immunohistochemical techniques with antibodies that selectively recognize tyrosine hydroxylase (TH), the initial and rate-limiting enzyme in catecholamine biosynthesis, and dopamine-β-hydroxylase (DBH), the enzyme that converts dopamine to norepinephrine. Although phenylethanolamine-n-methyltransferase, the enzyme that converts norepinephrine to epinephrine, has been detected in the neocortex (Burke et al., 1986), no adrenergic fibers have been visualized in the neocortex (Hokfelt et al., 1974). In addition, lesions of the locus coeruleus result in the loss of virtually all DBH-containing cortical axons (Lewis et al., 1988). Thus, DBH is considered to be a specific marker for noradrenergic fibers. In contrast, antibodies against TH would be expected to identify both dopaminergic and noradrenergic cortical axons. However, comparisons of DBH and TH immunoreactivity in monkey neocortex revealed the presence of DBH-labeled fibers in cortical areas and layers with little or no TH immunoreactivity, indicating the existence of DBH-containing, noradrenergic axons in which TH was not detectable immunohistochemically (Lewis et al., 1987, 1988; Campbell et al., 1987). In addition, there were multiple regions and layers of monkey neocortex where TH-immunoreactive fibers were very dense but DBH-labeled processes were sparse or nonexistent, demonstrating that the anti-TH antiserum used in these studies was sufficiently sensitive to be able to visualize a separate, non-noradrenergic population of cortical axons. These findings suggested that the anti-TH and anti-DBH antisera used in these studies labeled distinct populations of axons in monkey neocortex that appeared to be dopaminergic and noradrenergic, respectively. These differences in the populations of cortical axons labeled by the anti-TH and anti-DBH antisera were supported by findings in monkeys with histologically confirmed oblations of the ascending noradrenergic projections of the locus coeruleus (Lewis et al., 1988). In these animals, cortical DBH immunoreactivity was eliminated or at least substantially reduced, whereas TH immunoreactivity was never decreased compared to controls. Furthermore, dual label studies confirmed that distinct populations of cortical axons were labeled by the anti-TH and anti-DBH antisera (Noack and Lewis, 1989).

The dual label studies with anti-TH and anti-DBH antibodies also demonstrated that the selective labeling of cortical dopaminergic axons by anti-TH antibodies depends upon the anti-TH antibody employed. For example, although we found that some anti-TH antibodies did not label any DBH-containing cortical axons, other anti-TH antibodies did. However, the percentage of DBH-immunoreactive axons that were also identified by the latter anti-TH antibodies varied both with the anti-TH antibody employed and with the cortical region examined. For example, an affinity-purified polyclonal sheep antiserum raised against TH derived from PC12 cells labeled less than 20% of the DBH-immunoreactive axons in prefrontal regions but over 75% of those in visual regions (Noack and Lewis, 1989). These types of antibody and cortical region effects on the selective labeling of dopaminergic axons in primate neocortex have been confirmed by other investigators. Samson and colleagues (1990) also found that some anti-TH antibodies label only dopaminergic axons in monkey hippocampus, whereas Gaspar et al. (1989) reported that their anti-TH antiserum labeled only 10–15% of DBH-positive fibers in human prefrontal cortex and up to 50% of the DBH-containing fibers in primary sensory regions of human neocortex. These findings suggest that the selective labeling of dopaminergic cortical axons by some anti-TH antibodies might be due to differences in the affinities of anti-TH antibodies, and consequently, in their abilities to detect lower concentrations of TH. For example, biochemical studies have demonstrated that cortical dopaminergic fibers contain a much higher concentration of TH than do cortical noradrenergic fibers (Emson et al., 1978; Schmidt et al., 1979). Alternatively, the differential labeling of dopaminergic and noradrenergic cortical axons by anti-TH antibodies might be attributable to differences in the TH molecule in dopaminergic and noradrenergic fibers (Acheson et al., 1981; Joh and Reis, 1975; Haycock, 1991).

Finally, it should be noted that the ability of certain anti-TH antibodies to selectively label cortical dopaminergic axons in monkey neocortex has been supported by the findings of other investigators using different techniques. For example, the patterns of distribution of dopaminergic axons in monkey cerebral cortex revealed by the anti-TH antiserum used in our studies have also been demonstrated by Berger and colleagues (1988) using an autoradiographic technique based on the specific uptake of tritiated dopamine and by Goldman-Rakic and co-workers (1989) using anti-dopamine antibodies.

Distribution of catecholaminergic afferents in monkey prefrontal cortex

In studies using the anti-TH antibody that selectively labels cortical dopaminergic axons (Lewis et al., 1987), one of the most striking findings was the widespread distribution and regional heterogeneity of the dopaminergic innervation of monkey neocortex (Fig. 2). In contrast to the relatively restricted distribution of dopaminergic axons in rodent neocortex (see Berger et al., 1991, for review) dopaminergic axons were present in every cortical region of both New World squirrel and Old World cynomolgus monkeys. In addition, the labeled axons had a distinctive regional pattern of distribution which suggested a functional specialization of the dopaminergic innervation of monkey neocortex. For example, TH-containing axons preferentially innervated motor over sensory regions, sensory association over primary sensory regions and auditory association over visual association areas.

These types of regional differences in density of innervation, as well as differences in the laminar pattern of innervation, were also apparent across

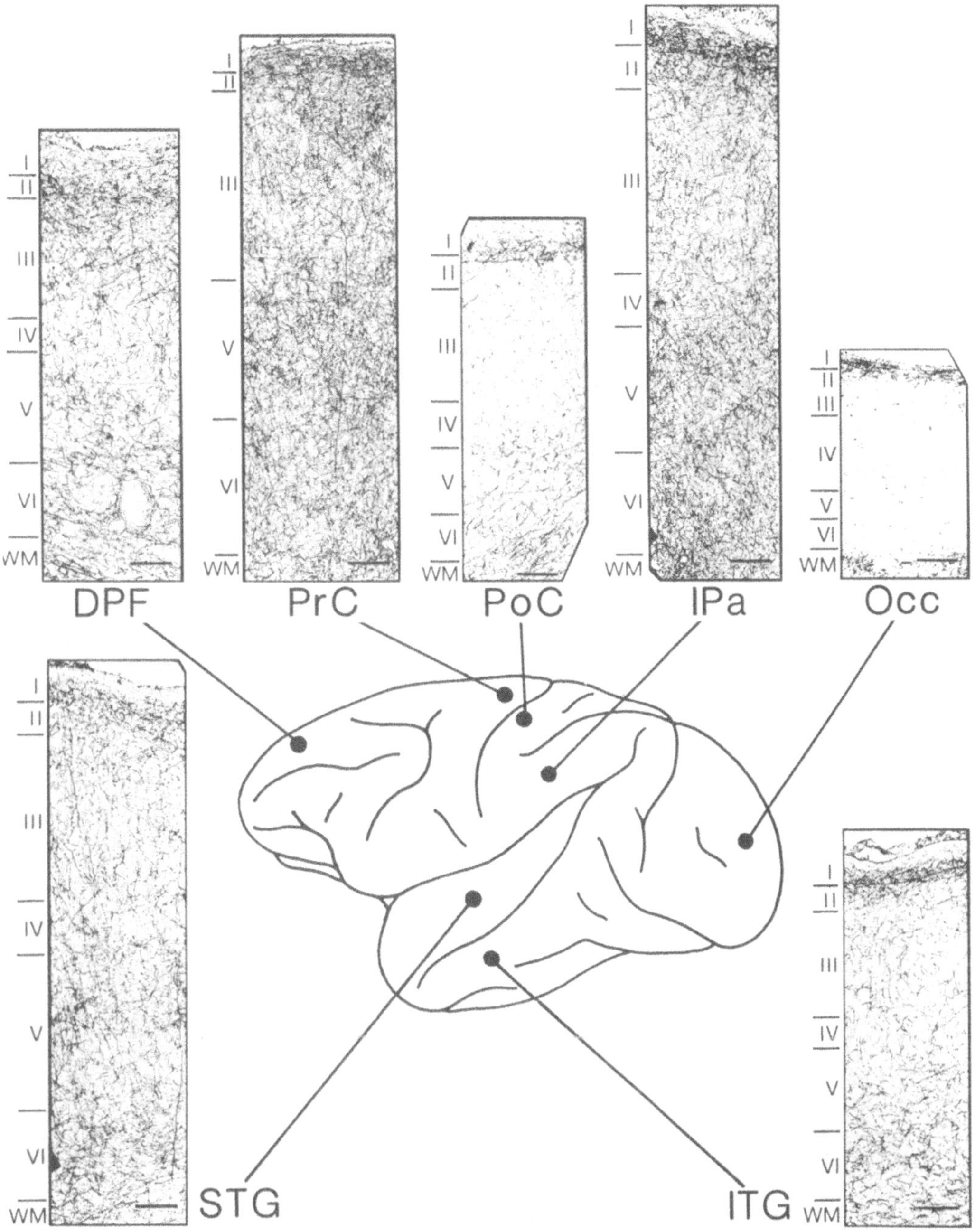

Fig. 2. TH-immunoreactivity in seven cytoarchitectonic regions of cynomolgus monkey cerebral cortex. Photographs are reversed images of darkfield photomontages. *DPF* dorsomedial prefrontal cortex (area 9); *PrC* precentral (primary motor cortex); *PoC* post-central (primary somatosensory cortex); *IPa* inferior parietal cortex (area 7); *Occ* occipital (primary visual cortex); *STG* rostral superior temporal gyrus (auditory association cortex); *ITG* rostral inferior temporal gyrus (visual association cortex). Roman numerals refer to the cortical layers and WM indicates white matter. Calibration bars equal 200 μm. (Reprinted from Lewis et al., 1987)

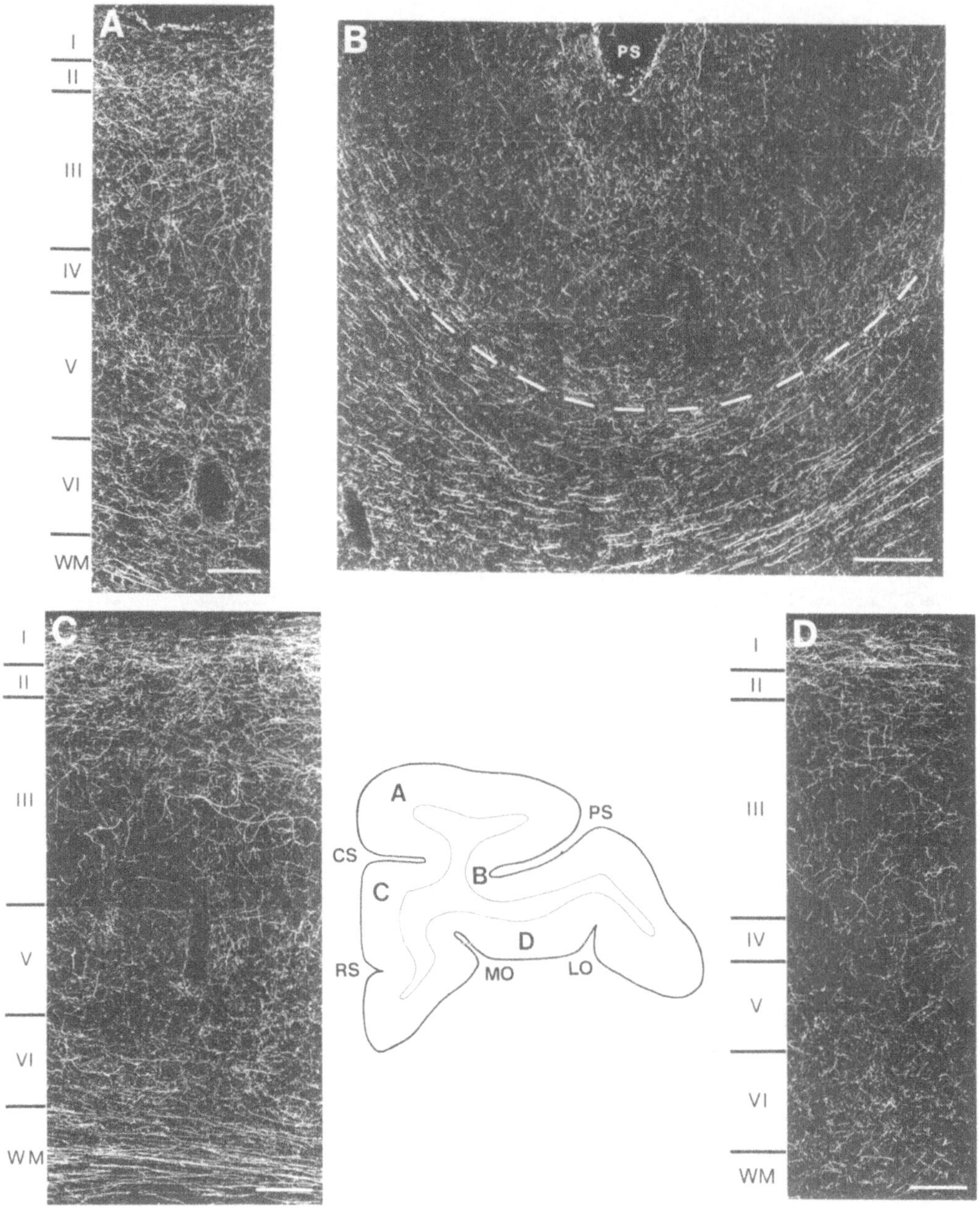

Fig. 3. Darkfield photomicrographs of TH immunoreactivity in cynomolgus monkey prefrontal cortex. The drawing in the lower center depicts a standardized coronal section of cynomolgus monkey prefrontal cortex. *CS* cingulate sulcus; *LO* lateral orbital sulcus; *MO* medial orbital sulcus; *PS* principal sulcus; *RS* rostral sulcus. The letters **A–D** indicate the approximate location of the photomontage with the corresponding letter. Calibration bars equal 200 μm in A, C and D, and 300 μm in B

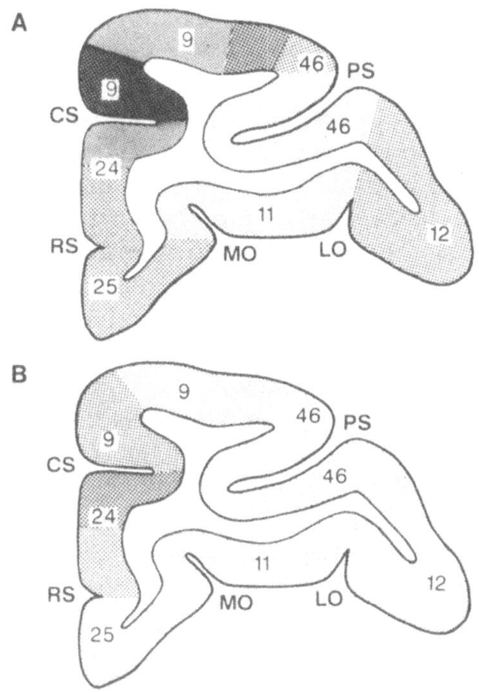

Fig. 4. Schematic representation of a coronal section from cynomolgus monkey prefrontal cortex indicating the relative density of TH-(**A**) and DBH-(**B**) immunoreactive fibers. Numbers refer to the cortical areas described by Walker (1940). *CS* cingulate sulcus; *LO* lateral orbital sulcus; *MO* medial orbital sulcus; *PS* principal sulcus; *RS* rostral sulcus

regions of monkey prefrontal cortex. In Old World cynomolgus monkeys, area 9 (Fig. 3A) had the greatest density of TH-labeled fibers, and the density was slightly lower in area 24 (Fig. 3C). Areas 11 (Fig. 3D), 12, 13 and 25 were intermediate in density and areas 10 and 46 (Fig. 3B) had the lowest density of immunoreactive fibers. These differences reflected gradients in density, both within and across cytoarchitectonic areas, on each surface of the prefrontal cortex. For example, on the dorsomedial convexity of prefrontal cortex, the density of TH-immunoreactive fibers increased in a rostral to caudal fashion. In addition, the medial surface had a ventral to dorsal gradient of increasing density. However, on the dorsal surface, the density gradient was more complex (Fig. 4A). In a medial to lateral direction, fiber density was greatest in area 9, decreased in the dorsal bank of the principal sulcus (area 46) and reached a nadir in the fundus of that sulcus. Fiber density increased again in the ventral bank of the principal sulcus and then increased slightly more in area 12 on the lateral convexity.

On a laminar basis, one of two general patterns of innervation were present. In most regions, labeled fibers had a bilaminar pattern of distribution, being most dense in layer I and the adjacent superficial layers, of intermediate density in layers deep V and VI, and quite low in density in the middle cortical layers. In contrast, the more densely innervated areas, such as area 9 tended to have labeled fibers more evenly distributed across all cortical layers.

Comparisons of these regional and laminar patterns of distribution of dopaminergic axons with those of noradrenergic axons in monkey prefrontal cortex revealed interesting similarities and differences (Lewis and Morrison, 1989). On a regional basis, DBH-immunoreactive, noradrenergic axons were most dense in areas 24 (Fig. 5C), and slightly lower in density in area 9 (Fig. 5A). Labeled fibers were intermediate in density in areas 11 (Fig. 5D), 12, 13 and 25 and low in density in areas 10 and 46 (Fig. 5B). This regional pattern of relative density was quite similar to that exhibited by the TH-positive dopaminergic axons. However, in comparison to dopaminergic axons, noradrenergic fibers had a substantially lower overall density and exhibited less marked regional differences (Fig. 4).

Despite regional differences in the density of labeled fibers, the laminar distribution of noradrenergic fibers was very similar across cytoarchitectonic areas of monkey prefrontal cortex. In every region, the density of labeled fibers was substantially greater in the deep cortical layers, especially layer V, than in the more superficial cortical laminae. Of these latter laminae, the deep portion of layer III and layer IV tended to have a greater density of noradrenergic fibers than the other layers. Thus, it appears that the two catecholaminergic systems tend to have complementary laminar patterns of innervation in monkey prefrontal cortex. That is, dopaminergic processes are most dense in the superficial layers, whereas noradrenergic axons are most dense in the infragranular layers.

In summary, the distinctive innervation patterns exhibited by TH-immunoreactive and DBH-immunoreactive axons reveal the regions and layers that may be the principal sites of action of dopamine and norepinephrine in exerting their effects on prefrontal cortical function. Comparison of the patterns of innervation of these two systems suggest that whether they share or have different target structures within a given prefrontal region, they may interact in regulating the output of that region. For example, of the pyramidal neurons that project to another cortical region, those in the superficial layers may receive dopaminergic input, whereas those in the deeper layers may be influenced by both dopaminergic and noradrenergic afferents (Fig. 6).

Distribution of catecholaminergic afferents in human prefrontal cortex

The potential importance of these patterns of innervation for understanding the role of prefrontal cortical catecholamines in the pathophysiology of schizophrenia depends, in part, upon the extent to which these studies in monkeys accurately predict the catecholaminergic innervation of human prefrontal cortex. In order to address this question, we compared the regional and laminar patterns of distribution of TH-containing axons in

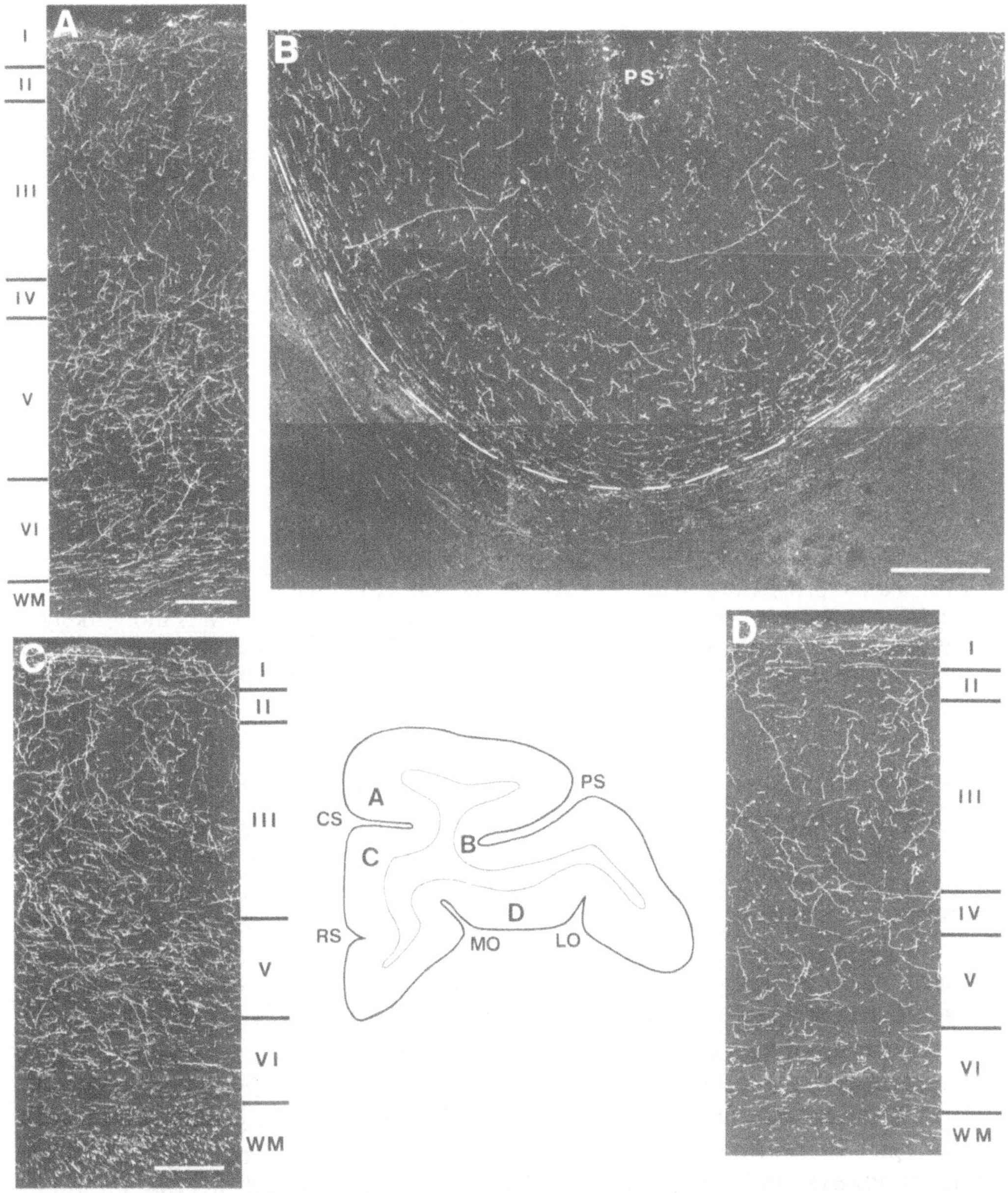

Fig. 5. Darkfield photomicrographs of DBH-immunoreactive axons in cynomolgus monkey prefrontal cortex. The drawing in the lower center is a schematic representation of a coronal section through prefrontal cortex. *CS* indicates cingulate sulcus; *LO* lateral orbital sulcus; *MO* medial orbital sulcus; *PS* principal sulcus; *RS* rostral sulcus. The letters **A–D** indicate the approximate location of the photomontage with the corresponding letter. Calibration bars equal 200 μm in **A**, **C** and **D**, and 300 μm in **B**

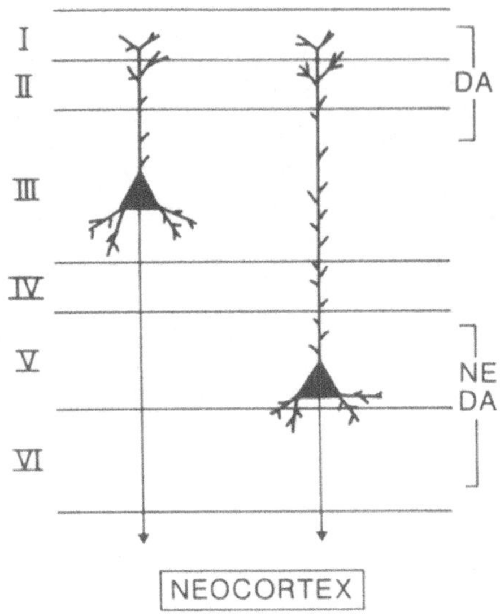

Fig. 6. Schematic drawing of the laminar location of two pyramidal neurons that furnish axonal projections to another cortical region. Note that the pyramidal neuron in layer III may receive dopaminergic input, whereas the one in layer V is more likely to be influenced by both dopaminergic and noradrenergic afferents. Roman numerals indicate the cortical layers. *DA* dopamine; *NE* norepinephrine

monkey and human prefrontal cortex. Both species showed very similar gradients across and within cortical regions in the density of labeled fibers. For example, on the dorsomedial convexity of the frontal lobes, fiber density increased in a rostral to caudal fashion from the frontal pole to the premotor and motor regions in both monkeys and humans. In addition, the medial cortical surface had a ventral to dorsal gradient of increasing fiber density in both species. The similarities between species in the distribution of TH-containing axons is further illustrated by comparison of Figs. 3 and 7. Note that on these coronal sections of prefrontal cortex, the density of labeled fibers was much greater in the dorsomedial prefrontal regions and anterior cingulate than in the lateral and orbital regions in both species. Although the relative regional differences in density were less striking, in both monkeys and humans, the cortical surface dorsal to the cingulate sulcus had a greater density of labeled fibers than the cingulate cortex located ventral to that sulcus.

Similarly, regional differences in the laminar distribution of labeled axons were also present in both species. The lightly innervated lateral and orbital regions of monkey and human prefrontal cortex both tended to have labeled fibers distributed in a bilaminar fashion, whereas the more densely innervated dorsomedial region had a more uniform laminar pattern of innervation. These comparisons indicate that the innervation of monkey prefrontal cortex by dopaminergic axons may be an accurate model of the innervation of human prefrontal cortex by that same system. These

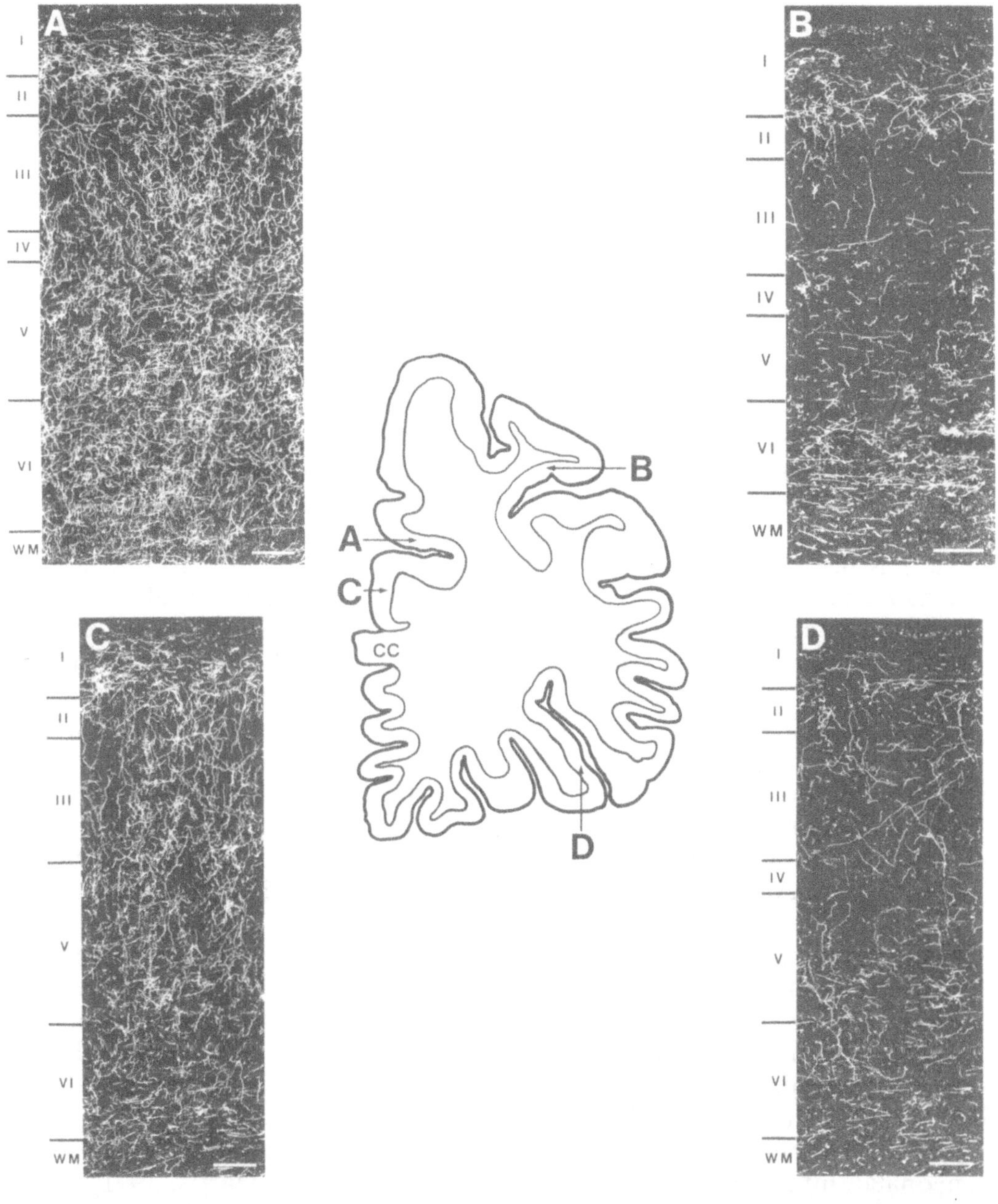

Fig. 7. Darkfield photomontages of TH immunoreactivity in human prefrontal cortex. The center panel is a camera lucida drawing of a section of human prefrontal cortex taken at the most anterior portion of the corpus callosum (*CC*). The letters **A–D** indicate the location of the surrounding photomontages. Note the differences in density and laminar distribution of the fibers present in medial prefrontal cortex (**A**) and anterior cingulate (**C**) than in the dorsolateral (**B**) and orbital (**D**) prefrontal regions. Calibration bars equal 200 μm in all panels

anatomical similarities suggest that the functional characteristics of the dopaminergic innervation of prefrontal cortex derived from studies in non-human primates may be reasonably extrapolated to the function of that system in human prefrontal cortex, which obviously can not be studied in the same fashion.

The influence of development on the dopaminergic innervation of monkey prefrontal cortex

In addition to emphasizing the roles of the prefrontal cortex and catecholamines in the pathophysiology of schizophrenia, a number of theories of schizophrenia suggest that it is a neurodevelopmental disorder. Observations on the typical age of onset, the season of birth and the presence of structural brain abnormalities in schizophrenic patients have all been considered to be consistent with a developmental origin of this disorder. For instance, in the case of structural brain abnormalities, increased ventricular size has been shown to predate the onset of psychotic symptoms, and to remain stable during the course of the illness (see Cazzullo et al., 1989, for review). In addition, the absence of gliosis in post-mortem studies demonstrating decreased tissue volume or cytoarchitectural disturbances in schizophrenic patients may also suggest a pathological process that occurs early in development (Roberts et al., 1986). Although a number of studies have investigated the development of the dopaminergic innervation of rodent prefrontal cortex, relatively little is known about the maturation of this system in monkey prefrontal cortex.

Biochemical studies have demonstrated that the dopaminergic innervation of monkey prefrontal cortex undergoes substantial change during postnatal development (Goldman-Rakic and Brown, 1982). Dopamine tissue concentrations in monkey prefrontal cortex appear to decrease during the first six months of life, and then gradually increase to adult levels. In contrast, catecholamine synthesis rates steadily increase during the same time frame. In order to determine if these measures of developmental shifts in the dopaminergic innervation of prefrontal cortex were associated with changes in the pattern of distribution of dopaminergic axons, we examined the distribution of TH-immunoreactive axons in prefrontal area 9 of neonatal, infant and adult rhesus monkeys (Lewis and Harris, 1991).

In adult rhesus monkeys, this prefrontal region contained a high density of TH-labeled fibers with a laminar distribution indistinguishable from that previously reported in the same cortical region of adult cynomolgus monkeys (Lewis et al., 1988). Numerous labeled fibers were present in all cortical layers, although layers deep I-superficial III were more densely innervated than the deeper layers. In contrast, the laminar distribution of TH-positive axons was different in three neonatal (4, 8 and 22 days of age) rhesus monkeys. In each of these animals, a bilaminar organization of labeled fibers was present in area 9 (Fig. 8A). Layers deep I-superficial III

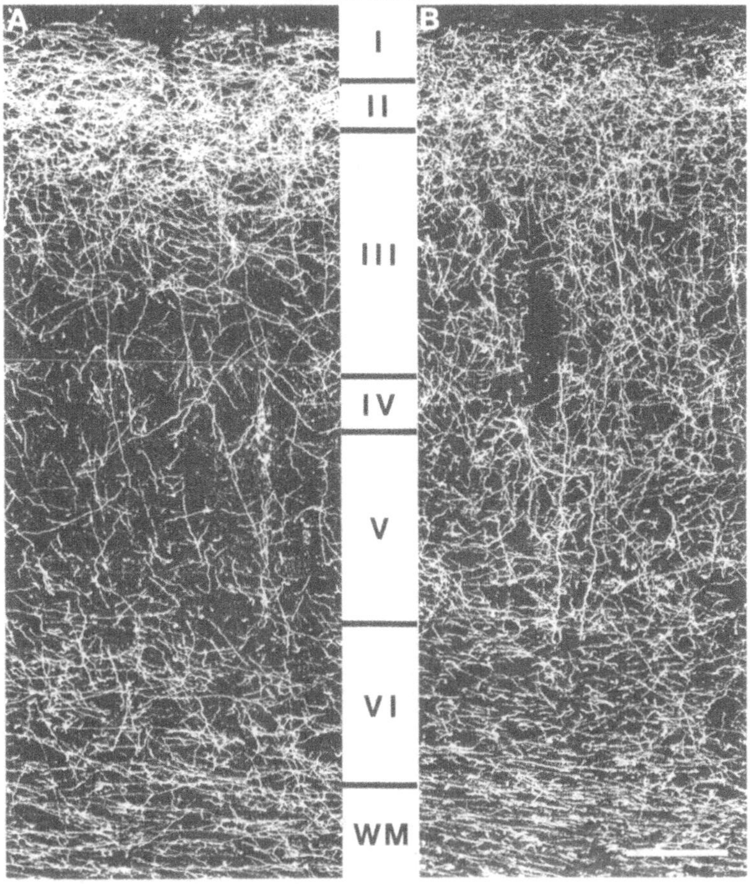

Fig. 8. Darkfield photomicrographs of TH-immunoreactive fibers in the dorsomedial convexity (rostral area 9) of rhesus monkey prefrontal cortex. Note the distinctive bilaminar distribution of labeled axons in the eight day old animal (**A**). In the 77 day old animal (**B**) the density of fibers has decreased in layers in *I-II* and substantially increased in layers deep *III-V*. Roman numerals indicate the cortical layers and WM indicates white matter. Calibration bar (200 μm) applies to **A** and **B**

contained a very high density of immunoreactive axons, layers deep III-superficial V had a much lower density and layers deep V-VI had an intermediate density of labeled fibers. Older infant animals (10–11 weeks of age) had a laminar pattern of innervation much more similar to adults than to neonatal animals (Fig. 8B).

Quantitative assessments of fiber density confirmed these marked differences between neonatal and adult rhesus monkeys in the laminar distribution of TH-containing axons in area 9 (Table 1). The greatest differences were evident in layers II and deep III. The density of labeled fibers in these areas shifted from a ratio of 3.3:1 in neonates to 1.2:1 in the adults. These differences appeared to be a consequence of two types of changes. Fiber density in the most superficial cortical layers declined, perhaps slightly

Table 1. Mean (±S.E.M.) density (per 0.01 mm^2) of TH-immunoreactive axons in area 9 of Rhesus monkey prefrontal cortex

Group	Age	n	Cortical layer* II[1]	III[2]	VI[3]
Neonates	4–22 d	3	65.3 ± 3.4^a	19.8 ± 1.0^a	28.0 ± 1.7^a
Infants	72–78 d	2	57.6 ± 4.9^a	38.0 ± 3.5^b	34.0 ± 1.0^{a,b}
Adults	4–17 y	3	47.8 ± 0.7^b	41.4 ± 1.2^b	34.8 ± 0.5^b

[1] ANOVA: $F_{2,5} = 17.50$; $p < .01$
[2] ANOVA: $F_{2,5} = 24.80$; $p < .005$
[3] ANOVA: $F_{2,5} = 9.39$; $p < .025$
* Within a column, values not sharing the same superscript are significantly different ($p < .05$)
Table reprinted from Lewis and Harris (1991)

during the first few months of life and more substantially between infancy and adulthood. In contrast, the deeper cortical layers, especially layers deep III through superficial V, exhibited a marked increase of TH-containing fibers during the first several months of life. The increased density of TH-positive axons in these layers appeared to persist unchanged into the adult, although fluctuations in fiber density during late infancy and adolescence can not be excluded.

If, as discussed above, these TH-immunoreactive cortical axons are dopaminergic, then the marked increase in number of axons in the middle cortical layers during the first months of life could be due to the arrival of new projections from the mesencephalon. Alternatively, these axons may have been present at birth but only achieved immunohistochemically detectable levels of TH as a consequence of some developmental change that influences local concentrations of TH. A third option is that the increased density of TH-positive axons in the middle cortical layers is due to the arborization and terminal proliferation of existing dopaminergic axons. It is interesting in this regard that the most striking increase in the density of TH-immunoreactive axons appears to correspond to the time of maximal synaptic density (2–3 months of age) in monkey prefrontal cortex (Rakic et al., 1986); however, in contrast to synaptic density, which subsequently declines to adult values, the TH-positive fibers in the middle cortical layers appear to persist into adulthood.

These types of developmental changes in the anatomical organization of dopaminergic afferents to monkey prefrontal cortex may be important determinants, or at least correlates, of a shifting role of dopamine in the functional circuitry of prefrontal cortex. Although it is not yet known whether similar types of changes occur in area 46 of prefrontal cortex, it is interesting that the time of apparent ingrowth of dopaminergic axons coincides with the appearance of the ability to perform delayed spatial tasks (Goldman-Rakic, 1987b), a behavior subserved by area 46 and which, in adults, appears to depend upon an intact dopaminergic innervation of this region (Brozoski et al., 1979). In addition, Morrison and colleagues (1990) have recently reported preliminary results suggesting that the density of TH-containing axons in layer III of area 9 is decreased in adult monkeys that were socially isolated from birth. These findings suggest that social isolation may have impaired the normal ingrowth of TH-containing axons to this region during the first months of development, and that this alteration in the normal adult pattern of dopaminergic axons in prefrontal cortex may possibly account for some of the behavioral abnormalities in social isolates. Finally, it should be noted that in adult monkey prefrontal cortex, TH-immunoreactive, dopaminergic terminals have been reported to form symmetric, presumably inhibitory, synapses on the somata, dendritic shafts and spines of pyramidal neurons (Goldman-Rakic et al., 1989). If pyramidal neurons are the major recipients of dopaminergic afferents to prefrontal cortex, then the increase in TH-containing axons in the middle layers of area 9 may reveal the potentiation of an inhibitory dopaminergic input to the pyramidal cells of these layers. That is, during the neonatal period, these pyramidal neurons may have received dopaminergic synapses predominately on their distal apical dendrites in the superficial cortical layers. However, despite a decline in the density of dopaminergic axons in the superficial layers, in adults dopamine may have a more potent inhibitory influence on pyramidal cell function through synapses on the proximal dendrites and somata of these neurons.

Intrinsic catecholaminergic systems in primate prefrontal cortex

The preceding sections have focussed on catecholaminergic axons that project to prefrontal cortex from the dopaminergic neurons in the ventral mesencephalon or the noradrenergic neurons of the locus coeruleus. However, recent studies have suggested that the neocortex may receive an intrinsic catecholaminergic innervation as well, based upon the observation that some cortical neurons are immunoreactive for TH, the rate limiting enzyme in catecholamine biosynthesis. The presence of these neurons in the cerebral cortex of nonprimates has been somewhat controversial. For example, TH-immunoreactive neurons have been reported to be present in adult hamster and rat cerebral cortex by some authors (Vincent et al., 1988; Kosaka et al., 1987), whereas other investigators have reported that TH immunoreactivity is detectable in neurons of rodent neocortex only transiently during postnatal development (Berger et al., 1985). However, cortical neurons labeled by polyclonal anti-TH antisera have been consistently found in post-mortem human neocortex (Kuljis et al., 1989; Trottier et al., 1989; Hornung et al., 1989; Gaspar et al., 1987). In our studies of human neocortex (Lewis et al., 1991), most TH-immunoreactive neurons had oval, vertically oriented cell bodies that gave rise to an ascending and descending dendrite (Fig. 9). Other neurons were horizontally arrayed,

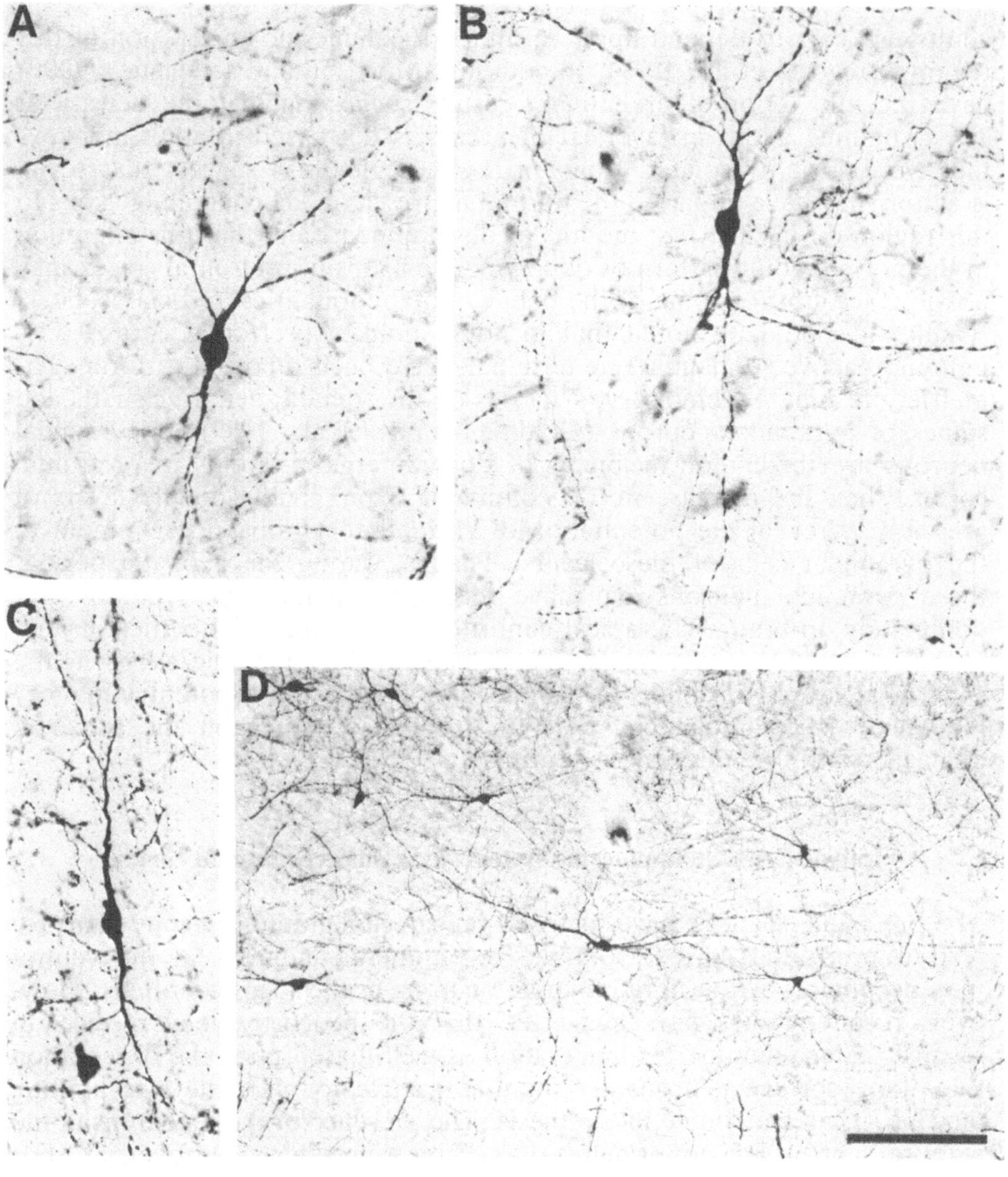

Fig. 9. TH-immunoreactive neurons in human prefrontal cortex. Note that many of these neurons have oval, vertically oriented cell bodies that give rise to dendrites that extend in both vertical and lateral directions (**A–C**). The highest density of these neurons is in layer VI (**D**). For all panels, the pial surface is toward the top. Calibration bar equals 80 μm in **A–C** and 200 μm in **D**

especially at the layer VI-white matter border where they were multipolar in form. Proximal dendrites were generally smooth but many distal dendrites had a very striking beaded appearance. Processes with an unequivocal axonal morphology have not been observed in our studies.

Table 2. Number of TH-immunoreactive neurons in human neocortex

B.A.[a]	Cortical region	Mean ±S.D.[b]
4	Primary motor	4.4 ± 2.6
3	Primary somatosensory	5.4 ± 3.0
17	Primary visual	2.0 ± 2.8
18	Visual association	7.4 ± 2.0
20	Higher order visual association	16.0 ± 1.6
9	Prefrontal association	14.5 ± 5.6
24	Paralimbic cortex	28.8 ± 6.6

[a] Brodmann's number
[b] Number of labeled neurons in a 4.0-mm-wide cortical traverse (n = 4)

Although these labeled neurons have been found in every cortical layer, the vast majority were present in layers V, VI and the subjacent white matter, with the greatest density in layer VI (Fig. 9D). On a regional basis, the density of labeled neurons was low in primary sensory and motor cortices, intermediate in sensory association areas, and greatest in higher order association and paralimbic regions (Table 2). These morphological characteristics and regional and laminar patterns of distribution are quite consistent with those reported by other investigators (Kuljis et al., 1989; Trottier et al., 1989; Hornung et al., 1989; Gaspar et al., 1987). Despite the general consistency of these observations, a number of questions regarding the anatomical and molecular characteristics of these neurons in human neocortex had not previously been addressed. For example, does the observed immunoreactivity actually represent TH, or is it due to a cross-reacting antigen recognized by polyclonal anti-TH antisera? Is the mRNA for TH expressed by human cortical neurons? If TH gene products are present in human neocortex, are these observations artifacts of the post-mortem state?

In order to address these questions, we conducted several different experiments (Lewis et al., 1991). First, to confirm that the cortical neurons labeled by polyclonal anti-TH antisera contained TH and not a cross-reacting antigen, identical immunohistochemical studies were conducted with two monoclonal anti-TH antibodies, known to recognize different portions of the TH molecule. Both antibodies labeled cortical neurons with a morphological appearance and with regional and laminar patterns of distribution similar to those seen with the polyclonal anti-TH antisera. Second, the expression of TH gene products by neurons intrinsic to neocortex was substantiated by in situ hybridization histochemistry using a riboprobe for TH mRNA derived from a human brain cDNA clone. These studies demonstrated the cellular localization of TH mRNA in cortical neurons that had laminar and regional patterns of distribution paralleling those of the

TH-immunoreactive neurons. For example, among the cortical regions examined, anterior cingulate cortex (area 24) contained the greatest density of grain clusters, whereas primary motor (area 4) and primary visual (area 17) cortices had the lowest number. In addition, grain clusters were most numerous in layer VI, the cortical layer containing the greatest number of TH-immunoreactive neurons. A recent study, using a probe to rat TH mRNA, also reported cellular localization of TH expression in human neocortex (Kuljis et al., 1989).

The presence of TH gene transcripts in human neocortex was further corroborated by polymerase chain reaction (PCR) methodology. Although the relative abundance of TH mRNA sequences in human neocortex were insufficient to be consistently detected by Northern analysis (Kaplan and Lewis, unpublished observations), Southern analysis of PCR products clearly revealed the presence of TH RNA sequences in human neocortex. In addition, these experiments demonstrated that the relative abundance of TH mRNA was greater in anterior cingulate cortex than in primary motor cortex, a finding consistent with the observed differences in the relative density of TH-immunoreactive neurons in these cortical regions. Other investigators have also used PCR analyses to show that alternatively spliced forms of TH mRNA are detectable in human cerebral cortex (Coker III et al., 1990).

Finally, TH-immunoreactive neurons (Kohler et al., 1983; Lewis et al., 1988) have been identified in perfused monkey brain, and in unperfused animals, their immunohistochemical detectability decreases with increasing post-mortem intervals. TH mRNA has also been detected in biopsied monkey neocortex (Kaplan and Lewis, unpublished observations). These findings indicate that the presence of TH gene products in human neocortex does not represent an agonal state or post-mortem artifact.

Thus, these findings demonstrate that TH gene products are expressed in intrinsic neurons in primate neocortex and consequently, they raise important questions regarding the role played by TH in cortical neurons. The reported absence of other catecholamine synthesizing enzymes (Gaspar et al., 1987; Trottier et al., 1989), such as aromatic L-amino acid decarboxylase and dopamine-β-hydroxylase, in human cortical neurons suggest that if the TH in cortical neurons is active, the only catecholamine that is synthesized is L-dihydroxyphenylalanine (L-DOPA). Based on similar anatomical evidence, other investigators have postulated that L-DOPA may be utilized as a neurotransmitter by some neurons (Hornung et al., 1989; Jaeger et al., 1984; Vincent et al., 1990). Indeed, DOPA immunoreactivity has been detected in neurons of the rat arcuate nucleus that contained TH but not dopamine or aromatic L-amino decarboxylase immunoreactivity (Okamura et al., 1988). In addition, L-DOPA has been reported to have direct synaptic actions (Bras et al., 1988), and endogenous L-DOPA may be released in a fashion similar to that of other neurotransmitters (Goshima et al., 1988). Since GABA has been colocalized with TH in cortical neurons (Trottier et al., 1989), L-DOPA may be utilized as a co-transmitter in these presumably inhibitory interneurons. Alternatively, L-DOPA released from intrinsic cortical neurons might serve as an additional source of precursor, through uptake mechanisms, for the catecholaminergic afferent axons that terminate in the infragranular layers of primate neocortex (Lewis et al., 1988, 1989; Levitt et al., 1984; Gaspar et al., 1989; Goldman-Rakic et al., 1989; Berger et al., 1988). This possibility might be plausible if catecholamines are, as some have suggested (Seguela et al., 1990), released nonsynaptically in the neocortex. Uptake of L-DOPA by catecholaminergic axons could provide a local mechanism for circumventing the limitation placed on catecholamine synthesis by the availability of TH and co-factor present in afferent axons.

Conclusions

The studies reviewed above indicate that the marked expansion and differentiation of the prefrontal cortex in primates is associated with an increase in the complexity of both the regional density and laminar distribution of catecholaminergic afferents compared to rodents. In addition, the innervation patterns of these systems in monkey prefrontal cortex appear to accurately predict those in human prefrontal cortex, suggesting that studies in nonhuman primates may be reasonably used to generate hypotheses about the nature of involvement of these systems in human neuropsychiatric disorders, such as schizophrenia. Furthermore, the distinctive and prolonged development of the dopaminergic innervation of primate prefrontal cortex and the possibility of an intrinsic catecholaminergic innervation of primate prefrontal cortex suggest new avenues of investigation into the roles of catecholamines in the normal and pathological function of human prefrontal cortex.

Acknowledgements

This work was supported by NIMH Research Scientist Development Award MH00519, NIMH grant MH43784 and NIMH Center for Neuroscience and Schizophrenia grant MH45156.

References

Acheson AL, Kapatos G, Zigmond MJ (1981) The effects of phosphorylating conditions on tyrosine hydroxylase activity are influenced by assay conditions and brain region. Life Sci 28: 1407–1420

Berger B, Verney C, Gaspar P, Febvret A (1985) Transient expression of tyrosine hydroxylase immunoreactivity in some neurons of the rat neocortex during postnatal development. Dev Brain Res 23: 141–144

Berger B, Trottier S, Verney C, Gaspar P, Alvarez C (1988) Regional and laminar distribution of the dopamine and serotonin innervation in the macaque cerebral cortex: a radioautographic study. J Comp Neurol 273: 99–119

Berger B, Gaspar P, Verney C (1991) Dopaminergic innervation of the cerebral cortex: unexpected differences between rodents and primates. TINS 14: 21–27

Bjorklund A, Divac I, Lindvall O (1978) Regional distribution of catecholamines in monkey cerebral cortex, evidence for a dopaminergic innervation of the primate prefrontal cortex. Neurosci Lett 7: 115–119

Bras H, Cavallari P, Jankowska E (1988) An investigation of local actions of ionophoretically applied DOPA in the spinal cord. Exp Brain Res 71: 447–449

Brown RM, Crane AM, Goldman PS (1979) Regional distribution of monoamines in the cerebral cortex and subcortical structures of the rhesus monkey: concentrations and in vivo synthesis rates. Brain Res 168: 133–150

Brozoski TJ, Brown RM, Rosvold HE, Goldman PS (1979) Cognitive deficit caused by regional depletion of dopamine in prefrontal cortex of rhesus monkeys. Science 205: 929–932

Burke WJ, Hanson DM, Chung HD (1986) A highly sensitive assay for phenylethanolamine N-methyltransferase in human brain. Proc Soc Exp Biol Med 181: 66–70

Campbell MJ, Lewis DA, Foote SL, Morrison JH (1987) The distribution of choline acetyltransferase-, serotonin-, dopamine-beta-hydroxylase-, and tyrosine hydroxylase-immunoreactive fibers in monkey primary auditory cortex. J Comp Neurol 261: 209–220

Cazzullo CL, Vita A, Sacchetti E (1989) Cerebral vertricular enlargement in schizophrenia: prevalence and correlates. In: Schultz SC, Tamminga CA (eds) Schizophrenia: scientific progress. Oxford University Press, New York, pp 195–206

Coker III GT, Studelska D, Harmon S, Burke W, O'Malley KL (1990) Analysis of tyrosine hydroxylase and insulin transcripts in human neuroendocrine tissues. Mol Brain Res 8: 93–98

Emson PC, Koob GF (1978) The origin and distribution of dopamine-containing afferents to the rat frontal cortex. Brain Res 142: 249–267

Fuster JM (1989) The prefrontal cortex: anatomy, physiology and neuropsychology of the frontal lobe, 2nd edn. Raven Press, New York

Gaspar P, Berger B, Febvret A, Vigny A, Krieger-Poulet M, Borri-Voltattorni C (1987) Tyrosine hydroxylase-immunoreactive neurons in the human cerebral cortex: a novel catecholaminergic group? Neurosci Lett 80: 257–262

Gaspar P, Berger B, Febvret A, Vigny A, Henry JP (1989) Catecholamine innervation of the human cerebral cortex as revealed by comparative immunohistochemistry of tyrosine hydroxylase and dopamine-beta-hydroxylase. J Comp Neurol 279: 249–271

Goldman-Rakic PS (1987a) Circuitry of primate prefrontal cortex and regulation of behavior by representational memory. In: Plum F, Montcastle V (eds) Handbook of physiology — the nervous system, vol 5. American Physiological Society, Bethesda MD, pp 373–417

Goldman-Rakic PS (1987b) Development of cortical circuitry and cognitive function. Child Dev 58: 601–622

Goldman-Rakic PS, Brown RM (1982) Postnatal development of monoamine content and synthesis in the cerebral cortex of rhesus monkeys. Dev Brain Res 4: 339–349

Goldman-Rakic PS, Leranth C, Williams SM, Mons N, Geffard M (1989) Dopamine synaptic complex with pyramidal neurons in primate cerebral cortex. Proc Natl Acad Sci USA 86: 9015–9019

Goshima Y, Kubo T, Misu Y (1988) Transmitter-like release of endogenous 3,4-dihydroxyphenylalanine from rat striatal slices. J Neurochem 50: 1725–1730

Haycock JW (1991) Four forms of tyrosine hydroxylase are present in human adrenal medulla. J Neurochem 56: 2139–2142
Hokfelt T, Fuxe K, Goldstein M, Johansson O (1974) Immunohistochemical evidence for the existence of adrenaline neurons in the rat brain. Brain Res 66: 235–251
Hornung JP, Tork I, DeTribolet N (1989) Morphology of tyrosine hydroxylase-immunoreactive neurons in the human cerebral cortex. Exp Brain Res 76: 12–20
Jaeger CB, Ruggiero DA, Albert VR, Park DM, Joh TH, Reis DJ (1984) Aromatic 1-amino acid decarboxylase in the rat brain: immunocytochemical localization in neurons of the brain stem. Neuroscience 11: 691–713
Joh TH, Reis DJ (1975) Different forms of tyrosine hydroxylase in central dopaminergic and noradrenergic neurons and sympathetic ganglia. Brain Res 85: 146–151
Kohler C, Everitt BJ, Pearson J, Goldstein M (1983) Immunohistochemical evidence for a new group of catecholamine-containing neurons in the basal forebrain of the monkey. Neurosci Lett 37: 161–166
Kosaka T, Hama K, Nagatsu I (1987) Tyrosine hydroxylase-immunoreactive intrinsic neurons in the rat cerebral cortex. Exp Brain Res 68: 393–405
Kuljis RO, Martin-Vasallo P, Peress NS (1989) Lewy bodies in tyrosine hydroxylase-synthesizing neurons of the human cerebral cortex. Neurosci Lett 106: 49–54
Levitt P, Rakic P, Goldman-Rakic P (1984) Region-specific distribution of catecholamine afferents in primate cerebral cortex: a fluorescence histochemical analysis. J Comp Neurol 227: 23–36
Lewis DA, Morrison JH (1989) The noradrenergic innervation of monkey prefrontal cortex: a dopamine-beta-hydroxylase immunohistochemical study. J Comp Neurol 282: 317–330
Lewis DA, Harris HW (1991) Differential laminar distribution of tyrosine hydroxylase-immunoreactive axons in infant and adult monkey prefrontal cortex. Neurosci Lett 125: 151–154
Lewis DA, Campbell MJ, Foote SL, Goldstein M, Morrison JH (1987) The distribution of tyrosine hydroxylase-immunoreactive fibers in primate neocortex is widespread but regionally specific. J Neurosci 7: 279–290
Lewis DA, Foote SL, Goldstein M, Morrison JH (1988) The dopaminergic innervation of monkey prefrontal cortex: a tyrosine hydroxylase immunohistochemical study. Brain Res 449: 225–243
Lewis DA, Melchitzky DS, Gioio A, Solomon Z, Kaplan BB (1991) Neuronal localization of tyrosine hydroxylase gene products in human neocortex. Mol Cell Neurosci 2: 228–234
Morrison JH, Hof PR, Janssen W, Bassett JL, Foote SL, Kraemer JW, McKinney WT (1990) Quantitative neuroanatomic analyses of cerebral cortex in rhesus monkeys from different rearing conditions. Soc Neurosci Abstr 16: 789
Noack HJ, Lewis DA (1989) Antibodies directed against tyrosine hydroxylase differentially recognize noradrenergic axons in monkey neocortex. Brain Res 500: 313–324
Okamura H, Kitahama K, Mons N, Ibata Y, Jouvet M, Geffard M (1988) L-DOPA-immunoreactive neurons in the rat hypothalamic tuberal region. Neurosci Lett 95: 42–46
Rakic P, Bourgeois J-P, Eckenhoff MF, Zecevic N, Goldman-Rakic PS (1986) Concurrent overproduction of synapses in diverse regions of the primate cerebral cortex. Science 232: 232–235
Roberts GW, Colter N, Lofthouse R, Bogerts B, Zech M, Crow TJ (1986) Gliosis in schizophrenia: a survey. Biol Psychiatry 21: 1043–1050
Samson Y, Wu JJ, Friedman AH, Davis JN (1990) Catecholaminergic innervation of the hippocampus in the cynomolgus monkey. J Comp Neurol 298: 250–263

Schmidt RH, Bhatnagar RK (1979) Assessment of the effects of neonatal subcutaneous 6-hydroxydopamine on noradrenergic and dopaminergic innervation of the cerebral cortex. Brain Res 166: 309–319

Seguela P, Watkins KC, Geffard M, Descarries L (1990) Noradrenaline axon terminals in adult rat neocortex: an immunocytochemical analysis in serial thin sections. Neuroscience 35: 249–264

Trottier S, Geffard M, Evrard B (1989) Co-localization of tyrosine hydroxylase and GABA immunoreactivities in human cortical neurons. Neurosci Lett 106: 76–82

Vincent SR (1988) Distributions of tyrosine hydroxylase-, depamine-beta-hydroxylase-, and phenylethanolamine -n- methyltransferase- immunoreactive neurons in the brain of the hamster (mesocricetus auratus). J Comp Neurol 268: 584–599

Vincent SR, Hope BT (1990) Tyrosine hydroxylase containing neurons lacking aromatic amino acid decarboxylase in the hamster brain. J Comp Neurol 295: 290–298

Walker AE (1940) A cytoarchitectural study of the prefrontal area of the macaque monkey. J Comp Neurol 73: 59–86

Weinberger DR, Berman KF, Zec RF (1986) Physiological dysfunction of dorsolateral prefrontal cortex in schizophrenia. I. Regional cerebral blood flow (rCBF) evidence. Arch Gen Psychiatry 43: 114–125

Weinberger DR, Berman KF, Illowsky BP (1988) Physiological dysfunction of dorsolateral prefrontal cortex in schizophrenia. Arch Gen Psychiatry 45: 609–615

Wyatt RJ (1985) The dopamine hypothesis: variations on a theme. In: Cancro R, Dean SR (eds) Research in the schizophrenic disorders. Spectrum Publications (The Stanley R. Dean Award Lectures, vol 2), pp 225–247

Author's address: Dr. D. A. Lewis, Departments of Psychiatry and Behavioral Neuroscience, University of Pittsburgh, 3811 O'Hara Street, Pittsburgh, PA 15213, U.S.A.

Subject Index

GPSR Compliance

The European Union's (EU) General Product Safety Regulation (GPSR) is a set of rules that requires consumer products to be safe and our obligations to ensure this.

If you have any concerns about our products, you can contact us on ProductSafety@springernature.com

In case Publisher is established outside the EU, the EU authorized representative is:

Springer Nature Customer Service Center GmbH
Europaplatz 3
69115 Heidelberg, Germany

Batch number: 10578403

Printed by Printforce, the Netherlands